Why Nothing Can Travel Faster than Light...

and Other Explorations in Nature's Curiosity Shop

Barry E. Zimmerman and David J. Zimmerman

CB
CONTEMPORARY BOOKS

Library of Congress Cataloging-in-Publication Data

Zimmerman, Barry E.
    Why nothing can travel faster than light—and other explorations
in nature's curiosity shop / Barry E. Zimmerman and David J.
Zimmerman.
      p.   cm.
    Includes index.
    ISBN 0-8092-3821-7 (pbk.)
    1. Science—Miscellanea.   I. Zimmerman, David J.   II. Title.
Q173.Z56   1993
500—dc20   93-27837
                                                     CIP

Cover design and illustration by Mark Anderson

20   19   18   17   16   15   14   13   12   11   10   9   8   7   6

To Marilyn and Sondra—
from their first suggestion that we write this book
to their proofreading of our final essays,
they have been a constant source of encouragement

To Amy, Tara, and Corie,
who are a joy and an inspiration

Contents

Acknowledgments

Thanks to William Rabus; Margaret Colvin; Carol Courtney; Jack Williams, of *USA Today*; Tanya L. Jones, of Alcor; and Alcor president, Steve Bridge, for their research assistance.

Introduction

"The universe is not only stranger than we imagine; it is stranger than we *can* imagine."
—J. B. S. Haldane

Strange and fascinating, to be sure. Science fact can be, and often is, more astounding than its fiction. Stephen Hawking made that evident in his bestselling book, *A Brief History of Time*. Hawking also made another point very clear: people are hungry for readable material about the world, the cosmos, in which they live. An abiding curiosity exists, whose glowing embers need only to be fanned.

The purpose of this book is to fan those embers. *Why Nothing Can Travel Faster than Light . . . and Other Explorations in Nature's Curiosity Shop* is a collection of thirty essays that answer some of the most intriguing questions in biology, chemistry, earth science, physics, and astronomy. It is not a book of trivia. It does not merely list amazing facts with perhaps a brief paragraph or two of explanation for each. There are a number of such books presently available. Neither is it a textbook. It does not seek to cover comprehensively only one area or field of scientific endeavor.

Instead, *Nature's Curiosity Shop* is a fast-paced, engaging treatment of many important and topical concepts and trends in science. Unrestrained by confining subject matter, it entertains as well as enlightens, moving from one phenomenon to another without belaboring a point.

Enough said. Sit back, turn the page, and enter the incredible, at times unbelievable, world of science.

Sky Wanderers– the Planets

If you look up into the sky on a dark and cloudless night, away from the city, you can see more than 3,000 points of light. These are the visible stars. Over a dozen evenings, you notice that several of them seem to move or change position among the other stars. Seen through a telescope, these moving stars are no longer points of light but disks. One even has a ring around it. Actually, they are not stars at all—they are *planets*. Stars always appear as points of light, even with the most powerful telescopes. They are too far away to be seen as anything else. And they are too far away to show movement from day to day. But planets do move. In fact, the word *planete* in Greek means "wanderer," and the planets are indeed sky wanderers. What have we learned about them over the centuries?

My Very Eager Mother Just Served Us Nine Pies. This sentence will help you remember the names of the nine planets in our Solar System as well as their order of increasing distance from the Sun: Mercury, Venus, Earth, Mars, Jupiter, Saturn, Uranus, Neptune, and Pluto.

Of the eight planets that circle the Sun (counterclockwise) along with Earth, five can be seen with the unaided eye: Mercury, Venus, Mars, Jupiter, and Saturn. They were discovered thousands of years ago by ancient astronomers who detected their movements across the sky. The other three planets—Uranus, Neptune, and Pluto—cannot be seen without the aid of a telescope.

Mercury, Venus, Earth, and Mars are sometimes called the terrestrial, or Earthlike, planets, because they are all relatively small, dense, and solid. You could land on them. Jupiter, Saturn, Uranus, and Neptune, on the other hand, are the gas giants—large and gaseous and not dense at all, like balloons of hydrogen and helium. (The terrestrial planets have very little hydrogen and helium.) You would drift right through them and eventually be crushed by their great pressure if you tried to land on them. They are called the Jovian, or Jupiterlike, planets. One planet, Pluto, is not classified at all. It is very small and far away and difficult to study. It is probably more terrestrial than Jovian.

Mercury

The Sun keeps Mercury on a very short tether. Being the closest planet to the Sun (less than half the distance of Earth from the Sun), Mercury is the most elusive. It is usually either hiding behind the Sun or so close to it that it can only be seen within an hour or two before sunrise or after sunset.

Mercury also moves around the Sun faster than all the other planets. In fact, in Roman mythology Mercury was the messenger of the gods because of his fleetness of foot. The planet's year is only eighty-eight days long, less than one-fourth of ours. Curiously, it spins much more slowly than Earth does, taking more than fifty-eight days for one complete turn.

Being a terrestrial planet, Mercury is also small and dense. It is the second smallest planet in the Solar System, less than half the diameter of Earth. In fact, it is smaller than three of the sixty-six moons that orbit the various planets. Unlike Earth, it has no moons and no atmosphere. Being so small, its low surface gravity cannot hold on to gaseous molecules for very long. A person weighing one hundred fifty pounds on Earth would tip the Mercurian scales at sixty pounds.

Close-up space exploration of Mercury by the *Mariner 10* probe in 1973 revealed several interesting features. It is highly cratered, like our moon. It also has a fairly strong magnetic field (though not as strong as Earth's). If you were lost on Mercury, you could use a compass to find your way home.

But could Mercury serve as home for space-settling earthlings? Not likely. Remember—Mercury has no atmosphere. It has no air to breathe. It also has no air to block out harmful radiation from space or to moderate temperature at the planet's surface. Mercury's temperature rises as high as 800° F (427° C) on the sunlit side, and drops as low as −280° F (−173° C) on the shadowed side. We're talking a difference of 1080° F (600° C)! On Earth, the greatest temperature difference is just over 180° F (100° C).

Venus

Named after the Roman goddess of beauty, Venus is by far the brightest point of light in the sky, about six times brighter than Jupiter and sixteen times brighter than Sirius, the brightest star in the heavens (excluding the Sun, of course).

Venus is sometimes referred to as the evening star and sometimes as the morning star because it generally rises within three hours before the sunrise and sets within three hours after the sunset. It also shows distinct phases, like our moon, when viewed through a telescope. Venus is also called our "sister" planet because it is nearly the same size as Earth and is Earth's nearest neighbor. Here the similarities end, however.

Venus is literally a furnace—by far the hottest planet in the Solar System. Lead would melt on its surface. But why the extreme heat? Venus is not nearly as close to the Sun as Mercury is; in fact, it's not that much closer to the Sun than Earth is. The answer, in a word, is *atmosphere*. Venus has a very thick, dense atmosphere—about one hundred times denser than Earth's. On its surface, the atmospheric pressure is nearly *one ton per square inch*! This is the same pressure that a diver would feel 3,000 feet below the ocean's surface. This atmosphere traps heat from the Sun and from the planet's interior, keeping the temperature constantly hot. (Venus's atmosphere, unlike Earth's, is rich in carbon dioxide, CO_2, a good heat-trapping gas.) It is a phenomenon known as the greenhouse effect. We have it here on Earth, too (see "Is Earth Getting Warmer?"), but on Venus it has run amok.

Perhaps because of its proximity to Earth, or perhaps because of its intriguing nature, Venus is the most visited planet in the Solar System. Only Earth's moon has been explored more frequently. From the nearly two dozen probes that have been sent to Venus, we have learned that Venus has mountains taller than Everest and depressions deeper than the Grand Canyon. It has no liquid water, no magnetic field, an atmosphere that rains acid, and—in all likelihood—no life. It takes about two-thirds of an Earth year for Venus to complete one spin—giving it the longest day of all the planets. It also spins clockwise, which is unusual.

Interesting fact: Because Venus has nearly equal periods of rotation and revolution, the sun hangs almost motionless in the Venusian sky and takes *nearly four years* to cross it!

Mars

The red planet. In ancient times it was thought to have gotten its color from the blood of warriors slain in battle. It has fired people's imaginations and been the subject of more science-fiction novels than any other planet. In fact, around the turn of the century an eminent American astronomer, Percival Lowell, suggested that the fine lines on the planet's surface were actually canals, or waterways, built by highly intelligent Martians to irrigate their crops and transport their goods on ships. This belief was still held by many as late as 1965, when close-up photographs from the *Mariner 4* probe showed no evidence of the fabled canals. It is no surprise, then, that Mars was the first planet to be orbited by a man-made satellite and the only one on which some form of life seems even remotely possible.

After a dozen years of space exploration (1964–1976) and a dozen trips to the red planet, we have learned a great deal. It is safe to say that there are no little green men navigating Martian canals. In fact, there are no canals. Mars is a dry planet—it has no liquid water. For this reason, we doubt that life of any kind, even single-celled bacteria, exists on Mars. Water did, however, run over the planet's surface at some time in its four-and-a-half-billion-year history. The channels that score its surface are erosion scars that indicate once-flowing rivers and streams. Perhaps there once was life on Mars. Studies of the Martian soil by *Viking* landers, however, have indicated no signs of life activity. In fact, scientists seriously doubt that life exists *anywhere* in the Solar System at the present time, except on Earth.

Mars is a colder planet than Earth, with a wider temperature range. At its equator, it *can* get as warm as a balmy spring day on Earth. It has a thin CO_2 atmosphere and polar caps of solid CO_2 that vaporize in summer and resolidify in winter. For this reason these caps, which appear white, fluctuate in size during the Martian year. Mars experiences a great deal of volcanic activity and has the largest volcano in the Solar System, Olympus Mons, which is three times taller than Mount Everest. It has essentially no magnetic field, so leave your compasses at home.

Mars revolves about the Sun at nearly half Earth's speed. A Martian year is equal to about two of ours. Mars has two moons, Phobos and Deimos. Its daytime sky is salmon colored, not blue like ours.

Mars is red in color not from the blood of slain warriors but because its surface is covered with rust.

Jupiter

King of the Gods! Truly a Goliath, Jupiter is more massive than all the other planets put together. More than 1,300 Earths can fit inside Jupiter. It is made up mainly of hydrogen and helium, like the Sun, and if it were only four to five times larger in diameter, it would have sufficient mass to fuse hydrogen to helium in its core, and there would then be two stars in our Solar System instead of one.

In 1610 Galileo pointed the newly invented telescope to the heavens and saw four bodies circling Jupiter. They have since come to be known as the Galilean moons. (One of them, Ganymede, is the largest moon in the Solar System.) In the 383 years since Galileo's observations, twelve additional moons have been found around Jupiter, as well as an elaborate ring system, a magnetic field 400 million times stronger than Earth's (all the gas giants have strong magnetic fields), and, of course, the Great Red Spot swirling at far greater than hurricane speeds in its atmosphere that just won't go away.

Jupiter was once thought to be primarily gaseous. It is still considered a gas giant, but due to extremely high pressure (more than 100 million times greater than Earth's atmospheric pressure) as one moves toward its center, gaseous hydrogen is changed to liquid hydrogen. It has a small, rocky core.

Next to Venus, Jupiter is the brightest "star" in the sky (with Mars close behind). Its day is about ten hours long. Because of Jupiter's great mass, it has a high surface gravity. On Jupiter you would weigh two and a half times as much as you do here on Earth.

Saturn

Lord of the Rings! Saturn is the only planet with a ring system easily seen from Earth with a small telescope. It is also the only planet that could float on water. It also has more moons than any other planet—twenty-three and still counting. The reason we're still counting is because of *Voyager*.

But what's *Voyager*? Probably the greatest space endeavor of all time. Two probes, *Voyager 1* and *2*, launched several weeks apart in 1977, encountered Jupiter in 1979 and Saturn in 1980. *Voyager 2* then continued alone to visit Uranus in 1986 and Neptune in 1989. With each *Voyager* encounter, the textbooks in planetary astronomy had to be rewritten. For example, only twelve of Saturn's moons had been discovered before *Voyager*. (Three of Jupiter's sixteen moons were also discovered by *Voyager*.) *Voyager* made it possible to closely study Saturn's atmosphere and magnetosphere. Several of Saturn's known rings were found to be composed of thousands of finer rings. Scientists learned more about Saturn and the other Jovian planets in the few years of *Voyager*'s grand tour than they had learned in the previous three and a half centuries since the invention of the telescope.

Next to Jupiter, Saturn is the largest planet, more than twice the size of Uranus or Neptune. Like Jupiter, it is composed mainly of hydrogen and helium gas, but because it is farther from the Sun than Jupiter, it is colder. Saturn takes about 29½ years to make one orbit around the Sun. Its day is about as long as Jupiter's.

Interesting fact: On Saturn you would weigh about as much as you do here.

Uranus

Uranus was the first planet to be officially "discovered."*
On March 13, 1781, an amateur astronomer, Sir William Herschel, noticed a fuzzy star in his 6.3-inch (16 cm) telescope. But stars aren't fuzzy—they're sharp points. Within four days, Herschel was able to detect motion of this object's movement—it was a *wanderer*. He had discovered the seventh planet in the Solar System.

Little more was learned about Uranus for more than two hundred years. It was known to have five moons, a bluish-green color, a period of revolution of eighty-four years, and a distance from the Sun about twice that of Saturn. Its temperature and rotational speed were not well established.

Then came *Voyager*. Ten additional moons were discovered, bringing the total to fifteen. A strong magnetic field

---

*The five planets already discussed, being visible to the unaided eye, had no exact *moment* of discovery. They were observed by prehistoric humans and recognized as planets by the ancients thousands of years ago.

was detected. A ring system was confirmed. Its composition and internal structure were accurately determined.

The unique feature of Uranus is its position: it is literally lying on its side, like a top that has stopped spinning. Scientists suspect that billions of years ago a large object slammed into Uranus, upsetting it so that it remained tilted 98° to the vertical. In other words, its axis is nearly horizontal, with its poles facing toward or away from the Sun. As a result, it has a "bull's-eye" appearance when viewed close-up, and day and night are each 42 years long. Because it is tilted *more than* 90°, it spins clockwise. All of the other planets except Venus spin counterclockwise.

The blue in Uranus's bluish-green appearance is not due to the scattering of blue light, as with Earth's atmosphere (see "Why Is the Sky Blue?"), but from the absorption of red light by the methane in its atmosphere.

Neptune

Ruler of the sea! Neptune and Uranus have often been viewed as twins among the planets. They are about the same size, have about the same surface temperature (colder than Saturn's), general composition, ring system, and period of rotation. There are differences, however. Neptune appears bluer, Uranus greener. Neptune is denser, has a much stormier atmosphere (with wind speeds *ten times* greater than Earth's most violent hurricanes), and does not lean on its side. It has eight moons (compared to Uranus's fifteen), six of which were discovered by *Voyager 2*.

Right now, Neptune is the farthest planet from the Sun. Because of the different shapes of Neptune's and Pluto's orbits, Pluto has been closer than Neptune to the Sun since 1979. (See Fig. 1.) But after 1999 Pluto will once again be the most distant planet from the Sun—for the next 228 years.

Figure 1

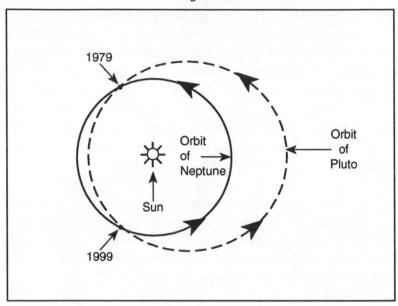

Pluto

God of the underworld! Pluto is the least understood of all the planets, the only one that has never been visited by a spacecraft. It is a small and lonely world. We're not even certain about its size, though we do know it is by far the smallest planet in the Solar System, less than half the size of Mercury. In fact, *seven moons are larger than Pluto, including our own*. Pluto has one moon of its own, called Charon.

Pluto is also by far the coldest planet. Its surface is most likely covered with a variety of ices. A thin, wispy methane atmosphere has been detected, which is unusual for a body so small. Our moon, after all, is larger than Pluto and has a greater surface gravity to hold an atmosphere, yet it has none. Why is this? Pluto's extreme coldness provides the explanation: the methane molecules move so slowly that they do not have enough energy to escape even the weak gravitation of Pluto.

Pluto takes longer to orbit the Sun than any other planet—245 years. You'd grow old before you were a third of the way around—or a third of a Plutonian year old!

Pluto is an oddball planet. All of the other planets orbit the Sun on about the same plane; Pluto's orbit is tilted 17°. All of the other planets orbit in nearly a circle; Pluto's orbit is egg shaped. These oddities hint that perhaps Pluto had an origin quite different from the other planets.

We have now passed the outermost planets* and are headed for the stars. Are there planets around these stars? We haven't seen any. With present technology, the reflected light that planets give off would be barely detectable across the distances between stars (although the Hubble Space Telescope will soon be searching for this reflected light). Wobbles in the motion of a few stars we have studied and fluctuations in the otherwise very regular radio signals of pulsar stars *do* indicate that there are planets moving around these stars (although only one instance has been confirmed). We certainly hope there are—for if there is life somewhere out there in space, it would not be on the stars themselves, which are far too hot, but on the dark, cool, invisible planets that move around them.

---

*Astronomers feel that there is a belt of small objects—miniplanets—*outside* the orbits of Neptune and Pluto; it is called the Kuiper Belt. One such miniplanet was tentatively discovered on August 30, 1992, and designated 1992 QB1. It is the most distant member of the Solar System, nearly twice as far out from the Sun as Neptune and Pluto currently are.

The Greatest Scientist of All

A number of years ago, fifty of the world's most respected scientists were asked to list their choices of the greatest scientists of all time. Of course, Albert Einstein appeared on most lists. So did Neils Bohr, who developed theories of quantum mechanics that helped determine the very structure of atoms. And Charles Darwin, who defined the mechanism of evolution. But only one name appeared on every list, at or near the top: a shy and retiring man by the name of Isaac Newton.

To quote Dr. Subrahmanyan Chandrasekhar of the University of Chicago, a Nobel Prize-winning astrophysicist: "It is fashionable today to think of Einstein as the epitome of scientific genius, and compared with us ordinary mortals, Einstein was indeed a giant. But compared with Newton, Einstein runs a very distant second." Einstein himself said that his own work would have been impossible without the discoveries of Newton.

According to Isaac Asimov, a popular and prolific American writer of science and science fiction, "Most historians of science would declare at once that Isaac Newton was the greatest scientific mind the world has ever known. He had his faults, heaven knows: he was a poor lecturer, something of a moral coward . . . and was at times subject to serious breakdowns. But as a *scientist*, he has no equal."

Newton was born in England in 1642, the year that Galileo died. His genius was not recognized early in his life. He received his bachelor's degree from Trinity College, at Cambridge, in 1665 and his master's in 1668, both without special distinction. Between 1665 and 1668 an epidemic of bubonic plague closed the University temporarily, and Newton went back home. Here, during moments of solitude and reflection, his genius blossomed. In a period of *eighteen months* between 1665 and 1667, Newton (1) discovered the laws of motion and universal gravitation, founding modern physics; (2) discovered the composition of light and the nature of color, founding modern optics; and (3) invented calculus, founding higher mathematics. Any *one* of these accomplishments would have placed him *among* the greatest scientists who ever lived.

Newton's laws of motion and universal gravitation, and how they apply to the physical world around us and to heavenly bodies in space, are described in a book he published in 1687, *Philosophiae Naturalis Principia Mathematica*, or *Mathematical Principles of Natural Philosophy* (also known simply as the *Principia*). The book was written in Latin and translated into English in 1729. It is considered to be the greatest single written work in the history of science. Bar none. To again quote Dr. Chandrasekhar: "In his *Principia*, Newton created the science of dynamics at a single stroke. The *Principia* underlies nearly every aspect of modern science." It was the climax of Newton's professional life. Another book, *Opticks*, published in 1704, describes his work with light and color.

In connection with his work on optics, Newton is credited with the invention of the reflecting telescope in 1668. The refracting telescope had been invented about sixty years earlier. It used lenses, which created a basic problem: when light passed through the lens, different colors that made up the light were bent differently and were focused at different points. This produced a blurred image.* Newton's reflecting telescope, on the other hand, used mirrors. Light does not pass through mirrors but bounces off them. Lights of all colors bounce the same way, so they focus at the same point, producing a sharp image. The reflecting telescope that Newton designed is still popular today and bears his name: the Newtonian reflector.

Halley's comet might also bear Newton's name, for without Newton, Halley would never have made his discovery. In the late 1600s Newton was applying his concepts of universal gravitation to celestial bodies in order to determine the motions of planets, moons, and comets. Newton's close friend and colleague Edmond Halley was intrigued by these calculations and visited Newton in 1684. His discussions with Newton led Halley to conclude that the bright comet of 1682 was the same one that had appeared in 1531 and 1607 and that it would again appear in 1758. Halley died sixteen years before he could test his prediction, but the comet did reappear on schedule, and it has been called Halley's comet ever since.

Halley recognized Newton's genius, and it was only with his support and encouragement that Newton wrote and

---

*The problem is called *chromatic aberration*. Today, lenses are made of a combination of different types of glass, which prevents chromatic aberration.

published the *Principia*. This was no small favor. According to Dr. Richard Westfall, a Newton biographer, had Newton died before writing the *Principia*, "We would at most mention him in brief paragraphs lamenting his failure to reach fulfillment." Through their association, each lent the other immortality.

Although the average person on the street might not readily identify Newton as the most brilliant scientist of all time, his greatness *has* been universally recognized. At the age of sixty, he was elected president of the Royal Society of London, a prestigious English scientific academy. Two years later, he was knighted by Queen Anne and became *Sir* Isaac Newton—the first British scientist to be so honored. Statues of Newton stand before the war memorial in Trinity College and at Westminster Abbey, where he is buried. His image has appeared on the British one-pound note, joining that of Shakespeare and the Duke of Wellington. A unit of force in the metric system bears his name—the newton. Four Newton postage stamps were issued in Britain in 1987 to honor the three hundredth anniversary of the publication of his *Principia*. And the Smithsonian Institution's National Museum of American History in Washington, D.C., presented a six-month exhibit devoted to the *Principia* in that same year.

Yet the greatest single work ever written in the history of science sells but several hundred copies each year. In fact, the only edition of the *Principia* in print in the United States is published in small numbers by the University of California Press—in a 1929 modern-English translation.

You must be wondering at this point if an apple really fell on Newton's head, providing him with sudden insight into universal gravitation? Is it fact or merely colorful myth? Well, an apple did fall—not on his head but near it. According to Newton himself, one fall day in Woolsthorpe during the plague, an apple from a nearby tree fell at his feet. This got him to thinking about gravity. He reasoned that the Earth's gravity is something that can act at a distance. After all, the apple wasn't touching the Earth, yet it was drawn to it by a force. Could not that *same* force that reached as far as the apple reach as far as the moon? The rest, as they say, is history.

Newton did have his faults, his greatness notwithstanding. He was a mean-spirited and vindictive man who spent most of his later life embroiled in heated disputes with other scholars. The most serious of these arose with the German philosopher and mathematician Gottfried Leibniz. Both he and Leibniz had developed—independently—calculus, an invaluable mathematical discipline. A bitter battle ensued over who had invented it first. (It is now clear that Newton invented it first but published it second.) Newton used his position as president of the Royal Society to publicly embarrass and humiliate Leibniz, accusing him officially of plagiarism. When Leibniz died, Newton is reported to have said that he took great satisfaction in "breaking Leibniz's heart."

Newton is thought to have suffered periods of temporary mental derangement. At least twice during his life he experienced bouts of depression and other abnormal behavior that bordered on psychosis. At the age of fifty he became quite ill, both mentally and physically. Several lines of evidence trace this aberrant behavior to his inhaling of mercury vapors. As an alchemist, he had a penchant for experimenting with mercury and its compounds.

Despite these "moments of madness," and a basically unpleasant and vengeful nature, Newton's genius remains unquestioned. Did he recognize his own greatness? He said of himself shortly before his death: "I do not know what I may appear to the world, but to myself I seem to have been only like a boy playing on the seashore, and diverting myself in now and then finding a smoother pebble or a prettier shell than ordinary, whilst the great ocean of truth lay all undiscovered before me."

When Newton died, at the age of eighty-four, his pall-bearers were two dukes, three earls, and the Lord Chancellor. Voltaire observed: "He was buried like a king who had done well by his subjects." No scientist before him had been so honored. Few after him were buried with such reverence.

Shortly after his death, the great eighteenth-century poet Alexander Pope summed up what the world thought of Newton:

Nature and Nature's Laws lay hid in Night.

God said, *Let Newton be!* and All was *Light*.

It remains Newton's epitaph.

Of Asteroids and Dinosaurs

Why did dinosaurs become extinct? This question has intrigued and perplexed both scientists and the general public for more than a century. Wild theories abound, some quite amusing. For instance, it has been proposed that dinosaurs died off because they became too massive to have sex—the male would crush his partner during mating. Another theory was that as plant life evolved, the diet of the dinosaurs changed, and they became fatally constipated. Yet another was that their eggs were all eaten by newly evolving and more agile and intelligent creatures—mammals. Some have suggested that new diseases wiped them out. A more credible theory takes evidence from air bubbles trapped in fossilized amber dating to the time of the dinosaurs. This air is 50 percent richer in oxygen than today's air. Perhaps as the oxygen level decreased to present-day standards, the dinosaurs suffocated.

Notwithstanding these compelling explanations, there

are generally two schools of thought regarding dinosaur extinction: (1) they died off suddenly, from some worldwide catastrophic event; or (2) they died off gradually, through a slow cooling of Earth's climate to which they could not adapt.

The catastrophists paint a very dramatic scenario: An asteroid or comet the size of Manhattan smashes into the planet at a speed of 15 miles (25 kilometers) per second, striking with an energy equal to 100 trillion tons of TNT (*five billion* times the power of the atomic bomb dropped on Hiroshima) and melting through rock as it burrows twenty-five miles beneath Earth's crust. In the aftermath of such an impact, thousands of tons of dust and debris are thrown into the atmosphere, enveloping Earth, blocking out sunlight, casting the entire planet in a deep dusklike darkness for months or years. In such a darkness, plants do not fare well—they need sunlight to make food. Animals that eat the plants do not fare well, either, nor do the animals that feed off them. The dinosaurs are among those animals that do not survive the darkness. In fact, according to some scientists, *60 percent* of the animal life on Earth do not survive the darkness.

But did the asteroid/comet crash really happen? The mass extinction of life on Earth is well documented, but was it caused by a monstrous collision? Let us follow the story from the beginning and find out.

A little more than 150 years ago, a British anatomist named Richard Owen coined the term *dinosaur* to represent a newly discovered group of organisms that had once lived on Earth. The word, taken from the Greek language, means "terrible" (*deinos*) "lizard" (*sauros*). At that time (1841) there were very few dinosaur fossils to study, and not much was known about them.

Our knowledge and understanding of the dinosaurs has

increased somewhat over the past century and a half. We know that they evolved about 225 million years ago from reptilian ancestors and that they died off about 65 million years ago. They varied greatly in size, ranging from less than two feet to more than 140 feet in length, and from twenty pounds to ninety tons. (An elephant, the largest extant land animal, weighs six tons.) They were classified as reptiles, which are cold-blooded and dim-witted creatures that lay eggs and are covered with scales.

Recently, however, the dim-witted, slow-paced view of dinosaurs has come into question. Studies suggest that many were quite intelligent and fleet-footed, as far as reptiles go. This view led Robert Bakker of the University of Colorado to propose in 1968 that dinosaurs were not cold-blooded at all but warm-blooded like ourselves. The argument is still not settled. It is likely that dinosaurs are "somewhere in the middle," to quote a group of British paleontologists. Or that the smaller, more active ones were warm-blooded, while the larger, lumbering beasts were cold-blooded. One scientist even suggested that the dinosaurs' metabolism may have varied by season: warm-blooded in winter and cold-blooded in summer.

So what happened to them about 65 million years ago that killed them off? Is there *any* evidence that an asteroid or comet smashed into Earth and did them in? Yes, there is.

In the late 1970s, a team led by the late physicist Luis Alvarez and his son, Walter Alvarez, of the University of California, Berkeley, made a startling discovery. They noticed that in layers of clay sediment about 65 million years old there was an abnormally high concentration of a certain element called *iridium*. Iridium is a heavy metal rarely found on Earth except deep beneath the crust, but it is present in fairly high amounts in comets and asteroids. At first, geologists thought that volcanic activity 65 million years ago may have thrown iridium into the atmosphere and that it settled and concentrated in a distinct layer. Ultimately, the facts did not bear out this theory.

How about an asteroid or a comet? The impact of such a body with Earth would vaporize much of the asteroid or comet, sending iridium-rich dust into the atmosphere world-wide. As the dust settled, it would form an iridium-rich layer of sediment. The Alvarez team found such deposits in Italy and Denmark. Since that time, such iridium-rich layers have been found at more than fifty additional sites scattered around the globe. That these layers are 65 million years old, coinciding with dinosaur extinction, is convincing evidence that an asteroid or comet killed the dinosaurs. Also, mineral deposits at certain sites show a type of fracturing caused by violent impact. This evidence is also consistent with the theory that an asteroid or comet collided with Earth, doing in the "terrible lizards." (Chunks of material in the Solar System that land on Earth are also called *meteorites*. A comet or asteroid, or fragment of such, that reaches the surface of Earth is a meteorite. As it hurtles through Earth's atmosphere before landing, it is called a *meteor*.)

If an asteroid or comet the size of Manhattan *did* strike Earth 65 million years ago, where is the hole, or crater, that it would have made? This has been a difficult question to answer. Different answers have, however, been put forward over the years. One was that the meteorite impact cracked or punctured Earth's crust, leading to volcanic activity that formed Iceland (which we know is of volcanic origin). How-ever, this theory is no longer widely accepted. The iridium layer in 65 million-year-old sediment is not that thick in Iceland. Geologists reason that the layer would be thickest around the site of impact, where the iridium was first hurled into the air. Also, along with the fine, whitish layer of iridium-rich clay sediment, there would be a second, coarser layer, which would be the result of the raining down of molten rock that splashed up from the impact, much as water in a pond splashes up when a pebble is tossed into it. This second layer, called the *ejecta* layer, would be present only within a radius of a few thousand miles of the impact site. It was not found in Iceland.

In 1990, Alan Hildebrand of the University of Arizona followed the iridium and ejecta layers to their thickest point. Two craters were found that were good candidates for the site of the meteorite collision. One was in the Colombia Basin, an area in the Caribbean Sea just north of Colombia. The other was on the northern edge of the Yucatán peninsula in eastern Mexico. Today the evidence overwhelmingly favors the Yucatán crater. It is just the right size—about 110 miles (180 kilometers) in diameter—shape, and composition. It is completely buried under rock and is situated on land, although 65 million years ago the area was under shallow ocean water. Believers of the meteorite-impact theory of dinosaur extinction feel that they have found not only the "smoking gun" but the bullet in the body as well.

One would think that at this point the riddle of dinosaur extinction has been solved. An iridium layer of clay proves a comet or asteroid struck Earth. It happened at about the right time. A crater was discovered to support the theory. However, doubts persist. So what's the problem?

The problem is whether or not the dinosaurs died out all that suddenly. An asteroid-impact scenario would require rapid extinction—within a few thousand years. Millions of years wouldn't do. Unfortunately, the fossil record does not give us a clear and indisputable picture. It is known that dinosaurs died out at *about* the same time—when one major geologic period, the Cretaceous, ended and another, the Tertiary, began. This transition is known as the Cretaceous-Tertiary (or K-T) boundary.

Peter Dodson, of the University of Pennsylvania, analyzed the fossil record of dinosaurs, focusing on that period of time. He concluded that eight million years before the end of the Cretaceous period, dinosaurs flourished. Six million years later, however, three-quarters of the existing genera were gone. This was bad news for the asteroid proponents. Many of the dinosaurs seemed to have died off before the asteroid/comet impact. However, the asteroid theorists point out that most of the dinosaur genera in North America survived and died off suddenly at the K-T boundary, when the meteorite struck. They also contend that counting genera is somewhat misleading. Many dinosaur genera had few members and were not well established in an evolutionary sense. The important genera, they say, and the vast majority of individual dinosaurs survived the late Cretaceous and died at the boundary, done in by a meteorite.

Over the past fifteen or so years since Luis Alvarez and his son made their startling discovery of iridium, the meteorite-impact theory has had a roller-coaster history. Conventional wisdom, at the moment, leans toward supporting the theory, especially with the discovery of the impact crater in the Yucatán. Indirect support of the theory also comes from the fact that there were at least *four* other instances of sudden mass extinctions on Earth over the past 450 million years. (Though asteroid-impact evidence from these periods, such as iridium-rich strata or telltale craters, is still forthcoming.) In fact, one that occurred about 240 million years ago pales the extinction of the dinosaurs, having decimated more than 90 percent of Earth's life forms. Ironically, it is believed that this mass extinction helped pave the way for the domination of the dinosaurs (which began to appear at about this time) by killing off most of their competition. In with a bang, out with a bang!

One final thought: If indeed the meteorite-impact theory is correct, it's probably a good thing that one did strike Earth and that the dinosaurs did die out. Otherwise the human race might never have evolved. With dinosaurs reigning supreme over the land and sea for countless millions of years, there was little room for the mammals. They remained small, shrewlike insectivores, unable to compete effectively. This all began to change with the "big bang."

Earth Without a Moon

Our moon, sometimes called Luna, is not a small or insignificant member of the Solar System. As far as moons go, it is surpassed in size by only four of the sixty-six known moons in the Solar System. It is actually larger than the planet Pluto (by *one and a half times*) and nearly as large as Mercury. Earth's diameter, in fact, is less than four times that of its moon. This size gives the moon a significant gravity. On the surface of the moon, you would weigh about one-sixth as much as you do on Earth.

With these impressive statistics, it is not surprising that the moon's presence is felt in many ways. What are some of these ways?

Tides

If you ever go to a seashore at different times of the day, you will notice that the water is not always at the same level. Twice in the course of a day the water level along the shore rises. At two other times during the day the water level recedes. This daily rising and falling of ocean waters is known as *tides*. The times when it reaches its highest point are called *high tides*. The times when it reaches its lowest point are called *low tides*.

The cycles between low and high tide vary in a regular and predictable way. They are caused chiefly by the gravitational effect of the moon on Earth. As Earth rotates, different portions of it face the moon. High tides occur in areas of the Earth that are directly toward or directly opposite the moon. In Figure 2, the Earth's oceans are represented by the shaded area. (The tidal effect is exaggerated.) Points 1 and 3 indicate areas of high tide. The tides are highest at these points because of the differential force of the moon's gravity on Earth's oceans. The moon is tugging hardest on the nearer water—point 1—and weakest on the farther water—point 3—causing the oceans at these points to bulge, or to be high. Low tides, on the other hand, occur in areas of the Earth that are neither facing the moon nor are directly opposite the moon but are halfway between these (points 2 and 4). In a period of twenty-four to twenty-five hours,

Figure 2

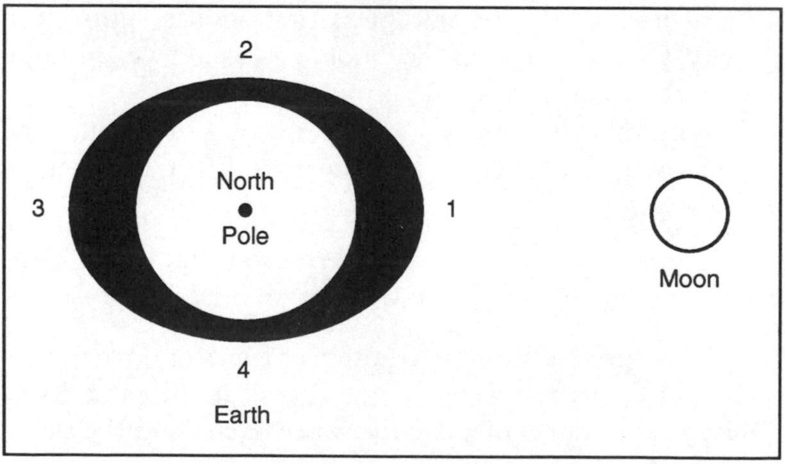

Earth will make one complete spin, causing two high tides and two low tides.

Without the moon, would there be any tides at all? Yes, but they would be much weaker. The Sun has an effect on the tides of Earth, but because of its greater distance from Earth, it is only half as strong as the moon's.* Without the moon, high tides wouldn't be very high, and low tides wouldn't be very low.

Tides vary seasonally, based on the position of the moon *and* Sun with respect to Earth. At certain times the Sun and moon are aligned with Earth. At these times, high tides are especially high and low tides are especially low. But again, without a moon the difference between high and low tides would not be nearly as great.

Life on Earth

The tides are important to the existence of many living things. The region of shoreline exposed at low tide and covered at high tide—called the *intertidal zone*—provides a unique environment suitable to a diverse group of plants and animals. Various worms and crustacea and seaweed thrive in this sometimes wet–sometimes dry environment; these, in turn, provide food for many bird species. This great diversity increases the competition for survival, which fuels adaptation and the evolution of new species. With a much narrower intertidal zone, as would be the case without a moon, there would be a great reduction in the diversity of organisms that exist on Earth.

It has been argued that life on land may never have evolved without a moon—or without an intertidal zone, to be more specific. Why is that?

Life on Earth began—more than 3.5 billion years

---

*The planets, especially Jupiter, also have an effect on the tides, but it is negligible.

ago—in the seas. That it did not originate on land is easily explained. Land environments are much more hostile and less suited for life. The temperature range between day and night, and from season to season, is much greater on land than in water. The effect of gravity is much greater, causing a constant weight burden. The buoyancy of water helps reduce this burden. (This is why sea creatures can be much larger than land creatures. The elephant is the largest land animal, weighing in at 6 tons. The largest sea animal, the blue whale, weighs in at 160 tons—nearly twice as heavy as the heaviest dinosaurs that ever lived. If the blue whale lived on land, it would be too heavy to move, and its spine would crack under its own weight.) Chemical reactions that are essential to life generally require water, which dissolves the chemicals so that they can mix and react. Also, reproduction generally requires water. The sperm must swim to reach the egg. And neither can afford to dry out.

For these reasons, life evolved and flourished in the waterways of Earth. To move onto land would have been a very dangerous and difficult transition. The intertidal zone provided an interesting possibility: an environment that could serve as a bridge between the water and the land. It is believed that living things took advantage of this environment and crossed the bridge about half a billion years ago— more than three billion years after life evolved in the seas. Without the intertidal zone—caused by the tides and, by extension, the moon—who knows if the process of evolution would ever have made the transition onto land? The physical barriers might have proven too great. And without life on land, there could be no *human* life as we know it—only mermaids and mermen.

Length of a Day

The tidal effects of the moon play a significant role in not only the evolution of life on Earth, but also in Earth's motions. Earth rotates, or spins, on its axis. This rotation causes day and night and the diurnal motions of the Sun and stars across the sky. It takes, on average, twenty-four hours to make one complete rotation with respect to the Sun. But it has not always been this way. Earth's tides, constantly sloshing around, smashing into cliffs and running up and down the shore, create friction—a resistance to motion—that causes the spinning Earth to continually slow down. Earth's rotation is slowing nearly two milliseconds (.002 seconds) per century. The day is getting one second longer every 62,500 years. That doesn't sound like much— there's no need to turn your clocks back yet. But over a long period of time, those milliseconds add up.

Assuming that the rate of slowing has been constant, the length of a day one billion years ago would have been *less than twenty hours*. Three billion years ago it would have been close to half as long as a day today. Without the moon, this tidal-effect slowing would have been greatly reduced. In other words, a day might be only ten or twenty hours long today.

Aside from the fact that we would all have to go to bed a bit earlier, there are profound implications to having a shorter day. Living things have what is called a biological clock—an internal mechanism relating to the functioning of the organism, such as its sleeping and eating habits, when it hunts for food, when it warms itself in the Sun, and so on. It helps the organism to make best use of the day. The biological clocks of most organisms operate on a cycle of twenty-four to twenty-five hours. With a much shorter day, it is likely that biological clocks would be correspondingly shorter.

As days became longer over millions of years, certain populations of organisms probably did not adapt well. Their biological clocks got out of sync with the day-night cycle, decreasing their survival advantage. This may have been the case with the dinosaurs, causing their gradual decline before the mass extinction of 65 million years ago.

A shorter day would have other intriguing consequences. For example, the strength and direction of Earth's winds would be different. Winds are affected by rotation; a faster spin would result in stronger winds. Hurricanes would blow harder and last longer. On Jupiter, which makes a complete rotation in less than ten hours, there are hurricanes that are many times more powerful than those on Earth, and they can last for decades or even centuries. (However, there are other factors that contribute to Jupiter's wind patterns.) With powerful winds constantly howling across Earth's surface, how might life have evolved differently? Perhaps trees would have much deeper roots, with few branches and leaves to catch the wind. How would we be different?

Cosmic Shield

If you look at the moon through a small telescope or pair of binoculars, you will see that it is not smooth looking but rather mottled and pitted. Many of these pits are craters that have resulted from the frequent bombardment by meteors. It is believed that without the moon to catch these meteors, many more would have collided with Earth instead. Some of these collisions may have been considerable. How might some of these have altered Earth's history and the evolution of life? Remember—it is believed that a meteorite impact some 65 million years ago led to the extinction of the dinosaurs!

Confusing the Issue

In 1543 the Polish astronomer Nicolaus Copernicus published a book in which he asserted: "At rest in the middle of everything is the Sun. . . . Thus, indeed, as though seated on a regal throne, the Sun governs the family of planets revolving around it." He is credited with establishing the concept that Earth revolves around the Sun (the heliocentric, or Sun-centered, theory). Until that time it had been generally accepted that the Sun (and planets and moon, for that matter) revolved around Earth (the geocentric, or Earth-centered, theory). This wrong theory had persisted for more than 2,000 years.

Why should people have been fooled for so long? In his essay "The Tragedy of the Moon," Isaac Asimov advances an interesting theory: our moon is the reason for the delay. In essence, Asimov maintains that the presence of a moon circling Earth confused the issue. It would have been difficult enough to understand that the movement of the Sun across the sky was actually due to Earth's movement and not the Sun's. But to comprehend at the same time the idea that the movement of the moon across the sky was due to a totally *different* kind of motion—that of the moon around Earth—was simply asking too much. This would have required the ancient astronomers to recognize two centers of revolution—the Sun for Earth, and Earth for the moon. It was much simpler to assume that both the Sun and moon revolved around Earth. Without a moon, celestial dynamics would have been less complicated, and we may have seen the truth more readily.

Tilt of the Axis

It has recently been postulated that the moon's rather significant gravitational pull on Earth helps to keep Earth's spin axis from wobbling or changing its position. It provides what is called a restraining torque. Presently, Earth's spin axis (the imaginary line passing through Earth from North

Pole to South Pole, around which it spins) is tilted 23.5° to its plane of revolution. Earth's tilt never varies more than 1.3° from this value. On the other hand, Mars, which does not have a large, stabilizing moon (it has two smaller ones), shows *much* greater variation in its spin axis. The axis is presently tilted 24° and has probably fluctuated from 10° to 50°. It is believed that without a moon, Earth's fluctuation would be at least as great.

What does all this mean? Well, for one thing, climate and seasons would be totally out of whack. A spin-axis tilt of 0° would mean no seasons at all. A tilt of 90° would mean six months of daylight followed by six months of darkness everywhere on Earth. At certain times polar ice caps would melt, raising the ocean levels as much as several hundred feet, flooding coastal areas. Ice ages might have had an entirely different history. (We've had at least six major ice ages as it is.) With these vastly different climatic situations, life on Earth (or certainly life as we know it) may never have happened.

So, without a moon life may never have moved onto land, humans may never have come to be, the day might be fourteen hours long instead of twenty-four, winds might be blowing with a force that makes hurricanes seem like summer breezes, and Earth might be spinning on its side, with snow at the equator.

But suppose we *had* evolved on an Earth without a moon? How might we have adapted to those much darker nights? Perhaps we would have six eyes instead of two—big, bulging eyes that were much more sensitive to light. Maybe they would be sensitive to infrared light (heat) as well as visible light. Perhaps . . .

Earth without a moon may yet come to pass. Our moon is moving away from us at the rate of a bit more than an inch a year. In five billion years it will be nearly 100,000 miles farther away than it is now. Its gravitational effect would be reduced by half. And Earth should be around for *at least* that long before the Sun calls it quits.

Why Is the Sky Blue?

Look up into the sky on a clear and cloudless day and you will see an expanse of pale blue stretching from horizon to horizon in all directions. But why blue? The light that reaches our world is a mix of *all* colors, not just blue. So why isn't our sky cherry red or mint green or flame orange? Why is the sky blue?

This question is often asked but not often explained. To better understand the answer, we must first examine the nature of light—in particular, sunlight, which is the source of light that illuminates our world and our sky. As we have already stated, this light is a rich mixture of colors, each representing a different *wavelength** of light. These colors cannot be separated and identified with the unaided eye but can be with a scientific instrument known as a spectroscope.

---

*Technically, it is frequency that determines color, but wavelength is often used as well and is better suited to this discussion. For a fuller explanation of wavelength and frequency, see "The Colors of Light."

When sunlight passes through a spectroscope, the colors that comprise it are separated, forming a band of colors called a spectrum—or *visible spectrum*, since it can be seen with the human eye. Each color occupies a specific place in the band, arranged as follows: red, orange, yellow, green, blue, indigo (deep blue), and violet. (High school science students are often taught to remember this sequence by memorizing the acronym ROY G BIV.) A rainbow is an example of a visible spectrum.

What happens when the mixture of colors in sunlight interacts with matter—molecules in the atmosphere, a glass of water, or the shirt or sweater you are wearing? Generally, matter can affect light in several ways:

1. Light can pass through the matter. When this happens, the light is said to be *transmitted*. Transparent media such as a pane of window glass transmit light well.
2. Light can bounce off the matter. When this happens, the light is said to be *reflected*. Most objects reflect light. An apple is red because its skin reflects red light. Chocolate is brown because it reflects brown light (that is, it reflects a combination of colors that the human eye sees as brown). In fact, the color of an object is the light that it reflects to your eye.
3. Light can be taken in by the matter. When this happens, the light is said to be *absorbed*. Absorbed light is not seen by the observer.

Imagine, now, a beam of sunlight traveling through the void of space toward Earth. As it enters Earth's atmosphere, it interacts with the molecules and bits of dust that make up the atmosphere. At this point something very interesting happens. Certain colors of light are selectively reflected, or bounced, off these molecules and dust particles in all directions. This process, in which light interacts with small particles and is reflected in all directions, is known as *scattering*. In the case of Earth's atmosphere, the color of light that is scattered the most is blue.

Why blue? The composition of the atmosphere determines the nature of the scattering. If the particles that do the scattering are smaller in size than the wavelengths of light scattered, blue will be the main color scattered. (Why this happens is rather complex and beyond the scope of our discussion.) This is the case with Earth's atmosphere: the molecules of gas and bits of dust are smaller than the wavelengths of light that make up a beam of sunlight. Blue light, therefore, is selectively scattered more than any color. In fact, it is scattered about ten times as much as red light. This selective scattering of blue light by very small particles is known as *Rayleigh scattering*.

To understand how Rayleigh scattering causes a continuously blue sky, look at Figure 3.

Figure 3

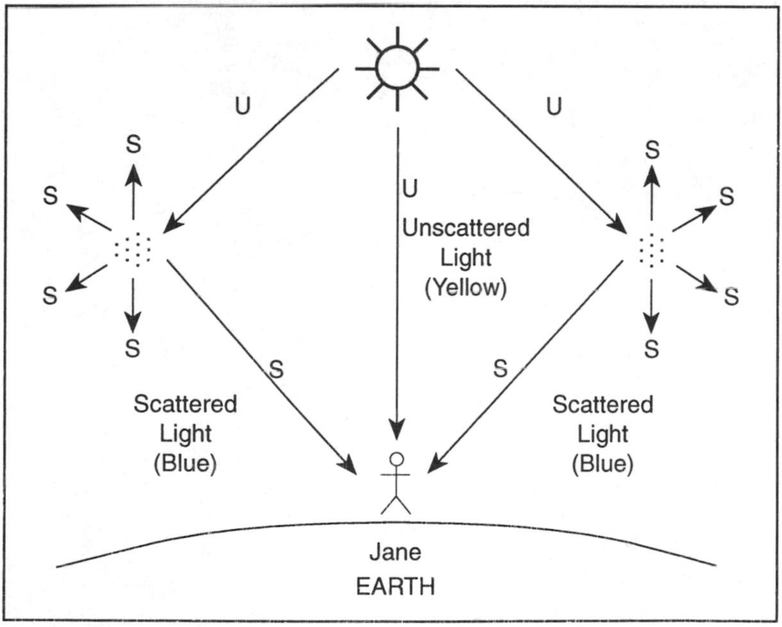

The three arrows labeled U designate rays of sunlight that contain a mix of all its colors (unscattered). This light is transmitted through the atmosphere. One of these rays is in the line of sight of the observer, Jane. She sees this mix of colors as yellow light. To Jane, or to anyone else who looks up at the Sun, it will appear yellow.

The arrows labeled S designate rays of sunlight that have interacted with the atmosphere and have been scattered. They are primarily wavelengths of blue light. If Jane looks anywhere but directly at the Sun, she will see this scattered blue light, as indicated by the S arrows in her line of sight. She will see blue sky.

The scattering of blue light by our atmosphere has other interesting effects aside from a pale-blue sky. Have you ever noticed that the color of the Sun changes as it moves in its daily course across the sky? A setting Sun is much redder than a noonday Sun. Why is this?

Figure 4

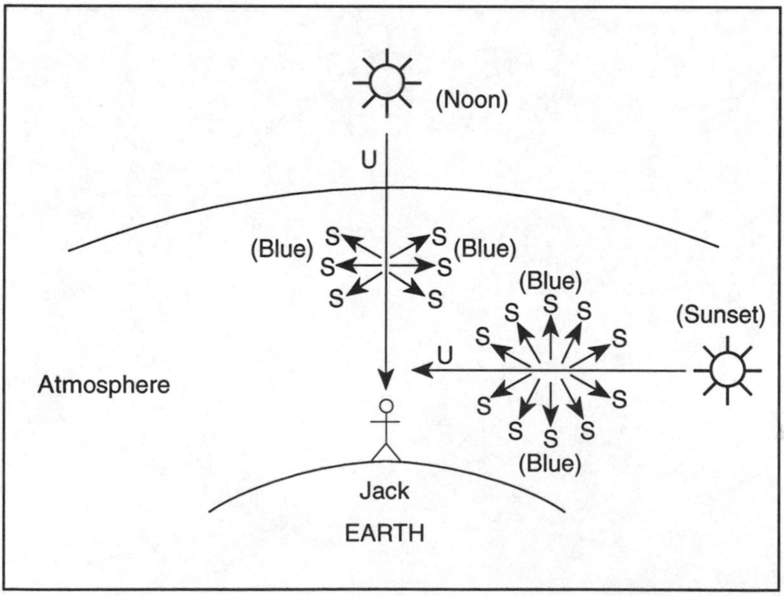

Figure 4 shows the Sun at two times—noon and sunset. Long arrows labeled U indicate rays of sunlight that are transmitted through the atmosphere and seen by Jack. As this light travels, it is scattered by the atmosphere. The short arrows labeled S indicate this scattering. As you already know, the scattered light is primarily blue. When blue light is scattered, it is removed from the ray of light (U), causing the ray to become redder. Since a setting Sun must pass through a greater thickness of atmosphere than a noonday sun, the light is scattered more and the sunlight is redder. You may be wondering if a Sun at sunrise is also redder than a noonday Sun. The answer is yes—for the same reason.

It seems that the atmosphere plays a major role in the coloring of our Sun and sky. But what if Earth had no atmosphere? Would the colors be very different? Absolutely! To better understand the differences, let us turn to a world that has no atmosphere—Earth's moon—and to someone who walked on that moon and looked up at the daytime sky. To quote Jack Schmitt of the lunar *Apollo 17* mission (one of six missions that landed on the moon in the late 1960s and early 1970s): "I had tried to anticipate what it would be like for many years. But there was no way to anticipate standing in the valley of Taurus-Littrow, seeing this brilliantly illuminated landscape with a brighter Sun than anyone had ever stood in before, with a blacker than black sky, and the mountains rising on either side." Why a black sky? Without scattering, the only light reaching the observer's eye would be directly from the Sun.

It is not necessary to travel all the way to the moon—about a quarter of a million miles—for this effect. Just hop in a space satellite and travel about a hundred miles up. At this altitude, the atmosphere is so thin that there is essentially no scattering. The sky appears black, the Sun appears more brilliant, and the moon, stars, and planets are all visible together. Wow!

We're not quite finished with blue skies. If Rayleigh scattering selectively scatters blue light, why aren't clouds also blue? If the particles doing the scattering are spherical in shape and *not* smaller than the wavelengths of light they interact with, all wavelengths are scattered equally; blue is not selected over any other color. This type of scattering, called *Mie scattering*, is what happens in a cloud. Clouds are mainly water droplets and ice crystals, which meet the size and shape requirements for Mie scattering. The result is a blend of all colors of light that comes across as white; hence, white clouds.

The thin, reddish atmosphere of Mars requires yet another explanation. It consists of many reddish dust particles that are too large for Rayleigh scattering and not of the proper dimensions for Mie scattering. These atmospheric particles simply reflect their own reddish color in all directions.

That's about it. We began by looking up and wondering why the sky is blue; then we scattered our way from a yellow Sun to red sunsets, through white clouds and black lunar vistas, and ended finally with red Martian skies. Why, I'm tickled pink!

The Colors of Light

Your radio alarm wakes you in the morning. You get up and prepare a cup of coffee in your microwave oven. Then you lie out in the backyard to get a suntan. Afterward, you visit your dentist for a checkup and get a series of X rays. At night you turn on a lamp and sit down to read the daily newspaper. Your chronic back problem is acting up, so you apply a heating pad to it.

What do these six activities have in common? In one form or another, they all involve the use of radiant energy, or *radiation*. What exactly *is* radiation, how is it produced, and what are some of its more important effects?

In a scientific context, radiation is defined as energy that is transmitted by electromagnetic waves—in other words, waves that have both an electric field and a magnetic field associated with them—hence, the more formal term *electromagnetic (e-m) radiation.**

*There is a particulate as well as a wavelike nature to e-m radiation, which is fundamental to quantum mechanics (see "What Is Quantum Theory?")

What causes certain devices such as lamps, microwave ovens, and x-ray machines to generate e-m radiation? Simply stated, e-m radiation is produced by any electrically charged particle that is accelerating or decelerating. These charged particles are generally electrons, which are present in all atoms and which move around the nuclei in distinct orbits. Their acceleration or deceleration occurs when they oscillate, or vibrate back and forth, as they jump from one orbit to another and back again. Electrons can be made to oscillate by absorbing energy. For example, when electricity is passed through a light bulb, electrons in the atoms of the bulb filament absorb the electricity, oscillate, and produce—as light (and some heat)—e-m radiation.

Radiation as electromagnetic waves was first suggested in 1864 by the Scottish physicist and mathematician James Clerk Maxwell. Drawing upon earlier experiments with electricity and magnetism by Ampère, Oersted, and Faraday, Maxwell was able to show that electric and magnetic fields act together to produce radiant energy, which, he predicted, traveled as waves. He also predicted the existence of radio waves. Twenty-three years later a German physicist, Heinrich Hertz, proved Maxwell correct on both counts by discovering radio waves. Ten years after that, X rays and gamma rays were discovered. It had become clear to the scientific community that there was a whole range of different types of e-m radiation, each with its own characteristics and importance to humankind. But *why* should this radiation differ, since it is all electromagnetic in nature and travels similarly through space as waves?

Nature of E-M Radiation

Electrons can oscillate at different rates, depending on the type of atoms they are part of and the amount of energy they are absorbing. Each oscillation produces an electromagnetic wave. Electrons in one group of atoms, for example, may oscillate 10 times/second, producing 10 waves/sec., while electrons in another group of atoms may oscillate 100 times/second, producing 100 waves/sec. The number of waves produced each second is called the *frequency* of the radiation. It is represented by the letter *f*. Since the waves travel outward from the source as soon as they are generated, frequency is also stated as the number of waves that pass a given point in one second. The term *hertz* (*Hz*), after Heinrich Hertz, is used to measure frequency. Each oscillation/sec. or wave/sec. is one hertz.

How fast do e-m waves travel outward? If they are not obstructed by matter in any way (as in a vacuum), they travel at a speed of about 300,000 kilometer/sec. (186,000 miles/sec.); that is, *all e-m radiation travels at this speed through empty space*. It is known as the speed of light. This presents an interesting situation. If the speed of all e-m radiation in a vacuum is 300,000 km/sec., then the distance the radiation travels in one second is 300,000 km. Yet the *number* of

Figure 5

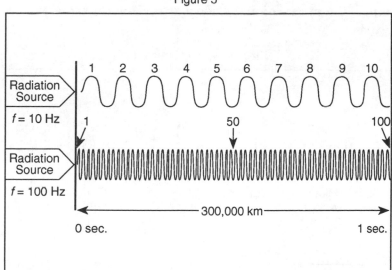

waves that travel that distance may differ. A source of radiation that has a frequency of 10 Hz will emit 10 waves/sec. Those 10 waves will cover a distance of 300,000 km in one second. On the other hand, if a source of radiation has a frequency of 100 Hz, it will produce 100 waves/sec., which will also cover a distance of 300,000 km in one second. Figure 5 illustrates these conditions. You will notice immediately that where fewer waves are produced per second (*lower frequency*), the waves are longer or more stretched out. The distance from one wave to the next is greater. This distance is known as *wavelength* and is represented by the Greek letter λ.

Thus, waves have two important components: frequency and wavelength. These components are oppositely, or inversely, related: as frequency increases, wavelength decreases, and vice versa. It is also these components that cause radiation to differ. Heat, light, and X rays are all basically the same—they are forms of e-m energy that travel as waves. But they differ very greatly from each other because of their different frequencies and wavelengths. The full range of e-m radiation is known as the *e-m spectrum*, which

Figure 6
Electromagnetic Spectrum

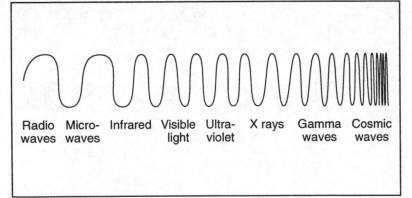

| Radio waves | Micro- waves | Infrared | Visible light | Ultra- violet | X rays | Gamma waves | Cosmic waves |

is illustrated in Figure 6. (The waves in Figure 6 are not drawn to scale. In actuality, cosmic rays may have wavelengths much shorter than a trillionth of a meter, while radio waves may be several thousand meters long. Visible light averages about half a millionth of a meter long.)

Changing Speed of Light

Frequency and wavelength of radiation can be combined in the following formula:

frequency $\times$ wavelength = velocity of radiation

or $f \times \lambda = v$

In a vacuum, the formula becomes $f \times \lambda = 300,000$ km/ sec. You can immediately see the inverse relationship between frequency and wavelength in these formulas. If velocity remains constant, then frequency and wavelength must be related in a reciprocal way.

But velocity does not have to remain constant. It can change if the medium in which the radiation is traveling changes. This makes sense. Can a bullet fired from a gun travel as fast through water as it can through air? Of course not. Generally, the denser the medium, the slower the radiation will travel through it. In this section we will limit our discussion to light energy and how it is affected by transparent media of different optical densities.

Visible light consists of a mixture of colors, each with a different wavelength and frequency. White light, or sunlight, contains *all* of the colors that make up visible light. When a ray of sunlight passes through a glass prism, these colors will separate and form a continuous band called the visible spectrum. The separation of light into its component colors is called *dispersion*. Why this happens will be discussed later in this section.

Figure 7

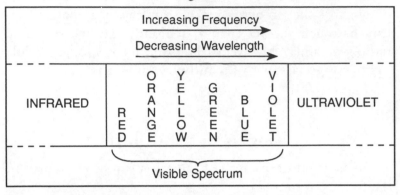

Figure 7 shows the continuous visible spectrum, produced by dispersion. You will notice that red has the lowest frequency and longest wavelength, while violet has the highest frequency and shortest wavelength. (See "Why Is the Sky Blue?" for a fuller discussion of the visible spectrum.)

Let us take a closer look at what happens when sunlight passes through a glass prism. In air, the sunlight moves quite rapidly—practically as fast as it does in a vacuum. As it enters the glass, a much denser medium, it slows down. Assume that it slows down to half its speed in air (a realistic assumption). The following formulas indicate this:

In air: $f \times \lambda = 300{,}000$ km/sec.
In glass: $f \times \lambda = 150{,}000$ km/sec.

If the speed of light changes, then the wavelength and/or frequency must also change. Scientists have determined that it is the wavelength that changes, *not* the frequency. If the speed is reduced to half, the wavelength of each color is also reduced to half; the frequency remains unchanged. The color of light also does not change. It is a function of frequency, not wavelength. (This applies to *all* e-m radiation, not just light.) Red is red because the source of radiation is sending out waves at a frequency of nearly

Figure 8
Refraction Through a Lens

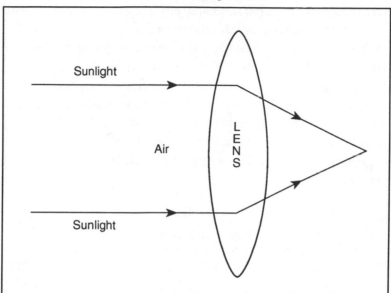

500 trillion/sec., not because the length of each wave might happen to be .65 millionths of a meter.

The slowing down or speeding up of light as it travels from one transparent medium to another causes the ray of light to bend or change its direction. This phenomenon is known as *refraction*. Figure 8 illustrates how refraction occurs as two rays of sunlight or white light pass through a lens. The property of refraction is what causes a lens to focus light. Without refraction, we would have no magnifying glasses, microscopes, telescopes, binoculars, or corrective lenses.

The story of refraction has an interesting twist to it. Earlier it was stated that a ray of light entering a glass prism slowed down from 300,000 km/sec. to 150,000 km/sec. This implies that all the colors in the light traveled at these speeds in the two media. That is not correct. Although all types of e-m radiation travel at the same speed in a vacuum, they do not in other transparent media. Generally, the higher the frequency of radiation, the more it changes speed. Within the visible spectrum, therefore, violet light changes speed the most and red the least. This difference in change of speed among the colors causes a difference in the amount each one bends by refraction. Because violet light changes speed the most, it is refracted the most. Red, on the other hand, changes speed the least and is refracted the least. Figure 9 shows how a lens refracts light. It acts much like a pair of prisms.

Figure 9

Refraction Through A Lens Showing Dispersion
(The Effect of Dispersion Is Greatly Exaggerated)

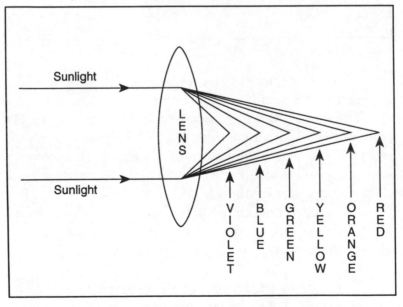

As you can see, the bending of light causes the colors that make up the light to separate. This, as already noted, is called dispersion, and it is a consequence of refraction.

Dispersion is both good and bad. With lenses, it creates a problem. The separated colors focus light in different planes, resulting in a blurred image. (This defect in a lens is called *chromatic aberration.*) It is the bane of telescope, microscope, and camera manufacturers. Compound lenses, which contain elements of different types of glass, help correct this problem.

With a prism, dispersion can be very useful. The separated colors help scientists identify chemical elements that have been heated and are giving off light. No two elements give off the same combination of colors or frequencies of light. Although the human eye cannot detect the different colors in the mixture, a prism can. An instrument that uses a prism for this purpose is the *spectroscope* (also discussed in "Why Is the Sky Blue?"). Using spectroscopes, astronomers have determined the composition of the Sun and other stars. In fact, the element helium was first discovered as a component of the Sun (*helios* in Greek means "sun"). The spectroscope has also been used to determine if stars and galaxies are moving toward or away from us and at what speed.

Doppler Effect

Have you ever listened to the whistle of a rapidly moving train as it rushes past you? The sound of the whistle appears to go down in pitch. It is higher as it moves toward you and lower, or deeper, as it moves away from you. This phenomenon is known as the *Doppler effect*, and it occurs with electromagnetic waves as well as sound waves.*

To understand the Doppler effect, imagine a duck sitting on a pond and flapping its wings at a set rate, say four

---

*Sound waves are not e-m waves. They are caused by the vibration of molecules, cannot travel through a vacuum, and are not electromagnetic in nature.

times per second. Each flap of the wings produces a wave in the water. They then travel outward in concentric circles, illustrated in Figure 10A. Waves will pass points A and B at a rate of four times per second. This is the frequency of the wave. A buoy floating at either point will bob up and down four times per second.

Figures 10A and 10B

Duck is stationary

Motion of duck

Now imagine that the duck is moving toward point A as it flaps its wings (Figure 10B). The distance between waves becomes shorter in the direction that the duck is swimming and longer in the opposite direction. As a result, waves approaching point A are shorter—and waves approaching point B are longer—than when the duck was stationary. Since all of the waves travel outward at the same speed once they are produced, not only will the waves be shorter at point A than point B, but more waves will pass point A each second. A buoy at point A will bob up and down more often than a buoy at point B. The wave frequency at point A will be higher than the wave frequency at point B.

The situation with e-m waves is analogous. Instead of a duck with flapping wings as the source of waves, imagine a light source, such as a star. Instead of buoys at points A and B, imagine observers with spectroscopes. As shown in Figure 11, the observer at point A will see waves that are shorter and of higher frequency than if the star were stationary. The observer at point B, on the other hand, will see waves that are longer and of lower frequency than if the star were stationary. Keeping in mind that red has the lowest frequency and violet the highest frequency of visible light (with blue close behind), the following are true:

Figure 11

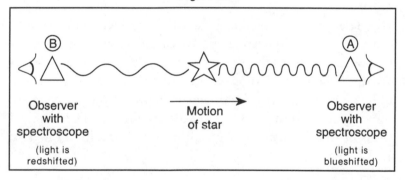

1. If frequency of light increases, the color of light shifts toward blue/violet. It is said to be *blueshifted*. (For some reason, scientists prefer blue to violet.) This will be the case if a star, or other source of light, is moving toward the observer or if the observer is moving toward the source. (It amounts to the same thing.) The faster the motion, the greater will be the blue shift.
2. If frequency of light decreases, the color of light shifts toward red. It is *redshifted*. This will be

the case if a star is moving away from the observer or if the observer is moving away from the star. The faster the motion, the greater will be the red shift.

A red shift does not mean that all light becomes red. It means that the frequency of radiation decreases. Blue light might become a bit greenish; yellow light might pick up a tint of orange. Red might move into infrared. The same is true with regard to a blue shift. It means that the frequency of radiation increases.

The Doppler effect—or *Doppler shift*, as it is also called—applies to all types of e-m radiation, not just visible light. It is the observed change in frequency (and wavelength) of *any* wave due to motion of the wave source or the observer.

The Doppler shift has its greatest application in astronomy. Using the spectroscope to calculate these shifts, astronomers have been able to determine if stars and galaxies are moving toward or away from us and at what speed. Curiously, virtually all galaxies except the very closest are moving away from us. They are all redshifted. The implications of this consistency are explored in detail in "How Did It All Begin?"

Wien's Law

If you take a metal wire and hold it in the flame of a gas burner, the wire will glow. It will become "red hot." As you continue to heat it, the glow will turn orange, then yellow, and finally white. Why is this happening?

It is happening because at different temperatures the wire emits radiant energy with different frequencies. *All* objects are continually giving off radiant energy in a mixture of frequencies. At low temperatures, however, these frequencies are low and lazy—too low to be seen as visible light. They are in the radio and microwave region of the spectrum. As temperature increases, so do the frequencies that the object gives off. At about 950° F (510° C), the object begins to emit the lowest frequencies we can see and begins to glow red. As frequencies increase, the object glows orange and then yellow. Eventually all the different wave frequencies to which the eye is sensitive are emitted, and we see the object as "white hot." At even higher temperatures— 30,000° F and greater, the object will dim as it glows increasingly into the ultraviolet region of the spectrum.

Wien's Law is a formula that relates temperature of an object to the principal kind of radiant energy it emits. It is used to determine the primary frequency or wavelength of energy emitted by different stars. According to Wien's law, the Sun emits mainly *green light*! Why, then, doesn't it appear green? (Hint: Remember that objects give off a *mix* of frequencies, or colors.)

The nature of light and how it interacts with matter are fascinating areas to explore. Through dispersion of light and spectroscopy we have discovered new elements and determined the temperature and composition of our Sun, the planets, and stars trillions of miles away. Through the Doppler effect we have learned that the universe is continually expanding, like a huge balloon. Through refraction, we have discovered magnification, bringing into focus the otherwise invisible worlds of the very small and the very distant. The list goes on.

Why Do Stars Shine?

And the earth was without form and void; and darkness was upon the face of the deep. . . .

And God said, Let there be light: and there was light.

To an inquiring mind, there is something unsatisfying about this passage from the Book of Genesis. How did God make the light? By what method was it produced? In other words, Why do stars, including our Sun, shine?

If you take a sheet of paper or piece of wood and burn it, light and heat are produced. If you take gasoline and spray it inside the cylinder of an engine, then produce a spark, it will also burn, releasing heat and light. In fact, the heat energy is released so rapidly that it causes an explosion. If you fill a balloon with hydrogen gas and place a lit match near it, the balloon will explode, producing light and heat and a loud noise. This is also burning—very rapid burning.

It just so happens that the Sun and all the other stars are made up primarily of hydrogen gas. Does this gas burn the same way in the stars as it does in a balloon filled with hydrogen and then ignited? Absolutely not! If it did, we would be in serious trouble. The Sun would produce less than *one billionth* as much energy as it does, and it would burn out much more quickly. Earth would be a frozen ball of ice, and there would be no life as we know it.

What's the difference, then, between star burning and ordinary burning?

To answer this question, we must go back to the atom. You may remember from basic high school science (see "How Dense Is Matter Inside a Black Hole?") that an atom has two main parts: (1) a small, densely packed center called a *nucleus*, which contains two types of particles, *protons* and *neutrons*; and (2) a larger but less dense region surrounding the nucleus, which contains a third type of particle, *electrons*. These electrons move around the nucleus in orbits, not unlike planets around the Sun. *Where* the electrons orbit is very important. Those that orbit close to the nucleus contain less energy. Those that orbit farther out from the nucleus contain more energy. In fact, the orbit in which an electron moves is called its energy level.

In ordinary chemical reactions, which include the burning of paper or wood or the exploding of gasoline or a balloon of hydrogen, only the electrons in the atom are involved—not the protons or neutrons. These electrons may go from one atom to another or be shared by two atoms. In doing so, they change their orbits or energy levels. Electrons that move to lower energy levels release energy, much as a ball rolling downhill releases energy. The rolling ball releases energy in the form of motion. Electrons release energy in the form of heat and light. This, in essence, is burning. During burning, the atoms of paper and wood or hydrogen combine with a gas in the atmosphere called oxygen. Electrons get rearranged, move to lower energy levels, and heat and light are released. Reactions in which electrons change energy levels are called *conventional chemical reactions*. Burning is a conventional chemical reaction.

Stars do not shine by conventional chemical reactions. First of all, there is essentially no oxygen in a star to combine with the hydrogen. Second, even if there were oxygen, conventional chemical reactions do not produce nearly enough energy. Electrons switching energy levels just doesn't get the job done. The stars require vastly greater amounts of energy—energy that can be found only in the nucleus. Such reactions are not conventional; they are called *nuclear reactions*.

Basically, there are two types of nuclear reactions: (1) those in which smaller nuclei combine to form a larger nucleus—called *nuclear fusion*; and (2) those in which a larger nucleus splits into smaller nuclei—called *nuclear fission*. Both types release huge amounts of energy. Fission occurs in nuclear reactors (power plants) and in atomic bombs, such as the ones dropped on Hiroshima and Nagasaki during World War II. Fusion is what takes place in hydrogen bombs, the Sun, and all other stars; it releases even more energy than fission. *It is nuclear fusion that causes stars to shine.*

But nuclear fusion does not happen easily. It requires extraordinarily high temperatures. How are stars able to reach these temperatures that can ignite their "fusion furnace," and cause them to shine?

Let us look at our Sun as a representative example. It formed about 4.6 billion years ago out of a huge cloud of gas and dust called a *nebula*. (The word means "mist" in Latin.) For some reason (possibly the close passing or violent explosion of a nearby star) the nebula began to contract, or condense. Once contraction was triggered, it continued

on its own due to the gravitational attraction of the matter
within the nebula. As the matter squeezed closer together,
pressure and temperature inside the nebula increased. This
squeezing together continued over many millions of years,
causing pressure and temperature to increase as well. At
some point the temperature became great enough to ignite
the nuclear furnace, and nuclear fusion began. It was then
that the Sun became a star.

Does the contraction of a star continue much beyond
the onset of nuclear fusion? No. The great amount of heat
produced by fusion causes an expansion or outward force
that just about counterbalances the inward force of gravity.
A delicate balance, or equilibrium, is created and the star
lives most of its life keeping this balance. It is said to be a
main sequence star during this time. More than 90 percent
of the stars in the cosmos, including the Sun, are main
sequence stars. As a star ages and runs out of its source of
fuel, hydrogen, it begins to lose its balance and moves off
the main sequence.

What is the minimum temperature required for nuclear
fusion to begin? Astronomers estimate it to be about
9,000,000° F (5,000,000° C or Kelvin*). Because hydrogen
is the fuel, the process is also called *hydrogen fusion*. It
involves a complex series of reactions that can be summa-
rized as follows:

4 hydrogen atoms combine to form 1 helium atom,
with the release of energy
or 4 H $\longrightarrow$ 1 He + Energy

The energy release is staggering. It does not come from

---

*The Celsius (C) and Kelvin (K) scales are not exactly the same, but
at high temperatures the difference is negligible. The Kelvin scale is
more frequently used in astronomy. With this scale, the degree
designation is omitted.

electrons moving to lower energy levels but from the conversion of matter *into* energy. One helium atom weighs a bit less than four hydrogen atoms. It has less mass. Where does the mass go? It is converted into energy. Einstein's famous equation, $E = mc^2$, predicts just how much energy. Each second, the Sun converts nearly *5 billion tons* of mass into energy. The conversion of merely *one gram* of mass (one millionth of a ton) into energy can light a million light bulbs for an hour. An area of Sun the size of a single-bed mattress produces more than 100 million watts of power per second. And the Sun is not even a particularly hot star.

With 5 billion tons of Sun being lost as energy every second and 600 billion tons being turned from hydrogen to helium, how long can it all last? The Sun must certainly run out of fuel soon. Remarkably, it has been going strong for over 4.5 billion years and is expected to continue doing so for at least another 5 billion years. The Sun is now entering middle age.

Do all stars shine for about ten billion years, as will the Sun, before their hydrogen engines run down? No—the life expectancy of stars varies greatly. Curiously, the smallest or least massive stars live the longest. They have masses about one-tenth that of the Sun and can shine for several trillion years. The downside is that they do not shine very brightly. The largest or most massive stars, on the other hand, which can be a hundred times more massive than the Sun, burn out in a million years or less. In the process, however, they shine very brightly—as bright as a million Suns.

More massive stars, in general, burn hotter, brighter, and use up their fuel much more rapidly than less massive stars. For example, a star twice as massive as the Sun will shine twenty times brighter and use up its hydrogen twenty times faster. It will burn itself out in a billion years. For this reason, it is believed that planets orbiting massive stars cannot have life on them. The stars do not shine long enough for life to evolve on their planets. It is estimated that life on Earth took at least 500 million years to come about.

Stars differ not only in their size and brightness but in the color of light they produce. In general, stars are divided into seven groups, or classes, designated by letters. These classes and important characteristics of each are listed in the following table. (A simple mnemonic may help you remember the letters of the star classes. Note the first letter of each word and not the sentence meaning! "O̲h, B̲e A̲ F̲ine G̲irl, K̲iss M̲e.")

Class	*Visible Color	Size (solar masses)	Average Surface Temperature (K)
O	blue-white	16–100	25,000–100,000
B	blue	2.5–16	20,000
A	white	1.6–2.5	10,000
F	yellow-white	1.05–1.6	7,000
G	yellow	.9–1.05	6,000
K	orange	.6–.9	5,000
M	red	.08–.6	3,000–3,500

*In reality, all star colors are merely tints, or faint deviations from white.

You may notice that the color of light a star gives off is based on its surface temperature. Its surface temperature, in turn, is based on its mass. The Sun is a class G star, with a surface temperature of about 6,000 K (10,300° F). More than 95 percent of all stars are less massive, cooler, redder, and dimmer than the Sun. Most stars, in fact, are type M. Nature likes its stars small.

What happens to stars when they run out of hydrogen? Depending on their mass, they take a variety of courses off the main sequence, but, in general, nuclear fusion continues with helium instead of hydrogen. Carbon forms as a result. Carbon can then fuse to form oxygen. In this way, many of the elements heavier than hydrogen or helium are formed— *elements that are necessary for life*. These elements are blown off into space in vast clouds as the star dies. Future stars that form from these clouds may then have planets that contain these heavier elements. Such stars are called second-generation stars. The Sun is a second-generation star.* First-generation stars, which form from the original material of the universe, are made up of only hydrogen and helium. (The universe was originally only hydrogen and helium.) Any planets that surround them are also made up of only hydrogen and helium. There can be no Earthlike planets orbiting a first-generation star—only gas giants like Jupiter or Saturn. There can also be no life on such planets.

On a clear, not-too-cold night, go outside and look up into the sky. You will see several thousand jewels handsomely displayed. Some will be bluish, while others will be cream-colored or pink. Some will be bright and twinkling, while others will be dimly visible. Yet for all their differences, they are very much alike. They are all like our Sun, turning matter into energy at a gluttonous rate and by the same remarkable process.

> And the earth was without form, and void;
> and darkness was upon the face of the deep.
> And God said, Let there be *hydrogen fusion*.

---

*Actually, it is believed that the Sun is the result of at least two previous star deaths. In this respect, it is a third- or perhaps fourth-generation star.

Measuring the Universe

The universe is at once infinitesimally small and infinitely large. The heads of human sperm cells, the smallest cells in the male human body, measure two millionths of a meter across. Ten thousand of them, placed next to one another, would cover one inch. Viruses are one hundred times smaller than these. The diameter of Earth, on the other hand, is 4 billion times larger than a human sperm cell. Would it be appropriate to use the same ruler to measure viruses and the size of planets? Put another way, would you measure the distance from New York to Los Angeles in inches or meters or how tall you are in miles or kilometers? Of course not. The unit of measurement must be appropriate to what you are measuring.

A Tale of Two Systems

When reading a textbook on science or traveling outside the United States, you most certainly have noticed that the units of measurement are different from those you are familiar with. Instead of ounces and pounds, you see grams and kilograms; instead of quarts and gallons, you see liters; instead of feet and miles, you see meters and kilometers. What's going on?

What's going on is two separate systems of measurement: the British, or Imperial, system and the metric system. As it happens, virtually the whole world, except for the United States, uses the metric system. Even the American scientific community uses the metric system exclusively.

Prefix	Basic Unit	Symbol	Value of Unit
exa	meter	Em	10^{18} m, or 1,000,000,000,000,000,000 m
peta	meter	Pm	10^{15} m, or 1,000,000,000,000,000 m
tera	meter	Tm	10^{12} m, or 1,000,000,000,000 m (trillion)
giga	meter	Gm	10^9 m, or 1,000,000,000 m (billion)
mega	meter	Mm	10^6 m, or 1,000,000 m (million)
kilo	meter	Km	10^3 m, or 1,000 m (thousand)
hecto	meter	hm	10^2 m, or 100 m (hundred)
deca	meter	dam	10^1 m, or 10 m (ten)
	meter	m	10^0 m, or 1 m
deci	meter	dm	10^{-1} m, or .1 m (tenth)
centi	meter	cm	10^{-2} m, or .01 m (hundredth)
milli	meter	mm	10^{-3} m, or .001 m (thousandth)
micro	meter	μm	10^{-6} m, or .000001 m (millionth)
nano	meter	nm	10^{-9} m, or .000000001 m (billionth)
pico	meter	pm	10^{-12} m, or .000000000001 m (trillionth)
femto	meter	fm	10^{-15} m, or .000000000000001 m
atto	meter	am	10^{-18} m, or .000000000000000001 m

Unfortunately, most people living in the United States are much more familiar with the British system.

The metric system was first established in France in the 1790s. In the 1950s the system was made uniform through international agreement and termed *Système international d'unités*, or International System of Units. For short, it is referred to as the SI. The table on page 66 lists the different units of length in the SI. Notice that the basic unit of length is the *meter*, or *m*. (One meter equals 3.28 feet, or 39.27 inches.) Other units of length are multiples of ten, larger or smaller than the meter. (One thousand meters—one kilometer—equals .62 miles.) Prefixes are used to help name these units.

One significant advantage of the SI is that it is extensive, including a range of units of length that can appropriately measure the very smallest and the very largest that the universe has to offer. The British system does not have nearly this range—there are inches, feet, yards, miles, and a small handful of additional units. Inches are far too large to measure atoms, and miles are far too small to measure galaxies.

Realm of the Small

To begin our exploration of size in the universe, let us start with the very smallest. How small can something be? Viruses are the smallest living things, but not the smallest *things*. Molecules are smaller than viruses. Atoms, which make up molecules, are smaller yet. Even smaller than atoms are the particles that make *them* up. That's where we'll begin.

Conventional models of atomic structure depict the atom as composed of protons, neutrons, and electrons. The protons and neutrons, particles that make up the nucleus of the atom, have diameters of about 100 femtometers, or one ten-trillionth of a meter (10^{-13} m). Exceedingly small indeed, but gargantuan compared to *strings*. Strings are one-dimensional, vibrating loops that, according to some particle physicists, underlie the structure of all matter and forces. They are about a billionth of a trillionth of a trillionth of an inch long, or 2.5×10^{-35} m. That's about four billion trillion times *smaller* than a proton or neutron. If strings do exist—and many reputable physicists believe they are only mathematical constructs that are not real—they are about as infinitesimal as things get.

But back to the atom.

Protons and neutrons, as already mentioned, have diameters of about one ten-trillionth of a meter, or 100 femtometers. About 10 billion of them can fit across the head of a pin. The nucleus of a typical hydrogen atom (smallest of all the elements) has only one proton, so its size is also about 100 femtometers. The largest atoms, however, have about 250 protons and neutrons combined, and the diameters of their nuclei are more than six times as large as a proton or neutron.

What about energy? Energy with wavelengths the size of atomic nuclei would be gamma rays. They have the shortest wavelengths and are the most penetrating of all types of electromagnetic energy.

So far we've discussed protons, neutrons, and nuclei, all of which make up atoms. How big are atoms themselves? They consist of much more than a nucleus. Far outside the central nucleus are shells, or orbits, in which electrons move. In fact, an atom is mostly empty space—the space between the nucleus and the orbiting electrons. The size of an atom would be the diameter of the orbit of its outermost electron. Small atoms have diameters about 1,000 times larger than a hydrogen nucleus, or proton. They measure in at 100 picometers, or 100 trillionths of a meter (10^{-10} m). Large atoms are about five times the size of these. This page is about one million middle-size atoms thick. Electromagnetic energy with wavelengths in this range would be X rays.

It's about time to mention another unit used to measure very small sizes; it is accepted among scientists worldwide but not previously discussed because it is not part of the *Système international*: the *angstrom unit* (Å). Very simply, it is 100 picometers, the size of small atoms. The largest atoms are 500 picometers, or 5 Å, across.

We're still in the realm of the micromicroscopic. As you know from basic science, atoms combine to form molecules. Molecules, in turn, combine to form larger structures, which in the living world may be viruses or cells. The most common cell found in human blood, red corpuscles, are disk shaped, with diameters just under 10 micrometers, or 10 millionths of a meter (10^{-5} m), or 100,000 Å. They are about fifty times smaller than a typical grain of sand. Wavelengths of energy this size are in the range of infrared radiation, or heat.

If cells are put together, they may form multicellular organisms. The smallest of these are about ten times larger than human red blood cells. Larger ones, such as spiders, start off about one hundred times the size of human red blood cells. That's a length of about 1 millimeter, or a thousandth of a meter (10^{-3} m), or 10,000,000 Å. At this size, we're moving outside the limit of the angstrom but into the range of British units. One millimeter is about $\frac{1}{25}$ inch. Energywise, this is in the range of microwaves. Yes, that's the type of energy used in microwave ovens to cook your food.

Multicellular organisms come much larger than spiders. The tallest human that ever lived was about 9 feet tall, or 2.75 meters. The largest living land animal, the elephant, measures nearly 33 feet, or 10 meters, from tip of trunk to tip of tail—about the width of a tennis court. Water dwellers come much larger. The blue whale, largest of *all* animals, measures more than 108 feet, or 33 meters, long. Plants can get even larger than blue whales. The tallest trees in the world, the redwoods in California, may grow as tall as 380 feet, or 116 meters—about four times as tall as the blue whale is long.

What type of energy has wavelengths this long? Radio waves range in length from the size of spiders to whales and redwoods—and longer. Radio waves include television as well as AM and FM radio transmissions.

Where do we go from here? There are no living things larger than whales and redwood trees—yet we are still near the beginning of our journey to measure the universe.

Realm of the Large

The tallest building in the world, the Sears Tower in Chicago, is 1,454 feet tall—443 meters. The tallest mountain in the world, Everest, measures about 8,850 meters, or 8.85 kilometers; this is about 5.5 miles. The world itself has a diameter of about 12,750 kilometers, or 12.75 megameters—7,923 miles.

For greater sizes or distances, we must move off Earth and begin to explore the Solar System—let us start with the Sun and move outward. At this point, the meter has become inadequate as a unit of measurement. Earth is, on the average, about 150,000,000 kilometers (93,000,000 miles), or 150 gigameters (1.5×10^{11} m) from the Sun. This distance is significant because it defines the *astronomical unit* (AU). Like the angstrom unit, the AU is not part of the *Système international*, but it is frequently used by the scientific community to measure distances between the planets.

If Earth averages 1 AU from the Sun (by definition), then Mars averages about 1.5 AU from the Sun, Jupiter about 5 AU, Saturn about 9.5 AU, Uranus about 19 AU, Neptune about 30 AU, and Pluto about 40 AU. The nearest star to the Sun is about 270,000 AU away. This is a distance of about 40,000,000,000,000 (40 trillion) kilometers, or 40 petameters (4×10^{16} m). It is also about 25,000,000,000,000 (25 trillion) miles.

In measuring between stars, even the AU becomes a bit cumbersome. SI units can always be used, of course, by attaching the appropriate prefix. Astronomers, however, have developed their own units to deal with the staggering distances they must routinely deal with. Two of these are the *light-year* (ly) and the *parsec* (pc). From this point on, distances and sizes are measured primarily in these units.

A light-year is, quite logically, the distance that light travels in one year (in a vacuum). Light travels about 186,000 miles, or 300,000 km, in *one second*. In one year, the distance would be about 5,900,000,000,000 miles— 9,500,000,000,000 km. This is more than 63,000 AU. The nearest star to the Sun is more than 4.2 light-years away.

The parsec is an even larger unit than the light-year. It is a contraction of the term *par*allax *sec*ond. Its derivation is complex and beyond the scope of our discussion. Suffice it to say that 1 pc = 3.26 ly = 204,265 AU. The nearest star is 1.3 parsecs away. Within our galaxy, the Milky Way, there are between 200 and 300 billion stars. They are arranged in a kind of spiral wheel, or disk, which measures about 100,000 ly or 30,000 pc across. The Milky Way, in turn, is part of a relatively small group of about 30 galaxies called the Local Group. The diameter of the Local Group, according to recent estimates, is between 4 and 5 million ly, or about 1.5 million pc. (The term *kiloparsec* is often used when distances reach this magnitude.)

Yet the Local Group is only a small sampling of galaxies in the universe. If you stretch out your arm and make a fist and point it *anywhere* in the night sky, it will cover an area that contains about a million galaxies. Most of these galaxies are arranged in groups, much as our Milky Way is part of the Local Group. The nearest galactic group to our Local Group is about 7 million ly (just over 2 million pc) away. Others have been detected that are billions of light-years away.

The farthest-out objects that astronomers have detected are called *quasars*. They were discovered in 1964. A quasar is enormously energetic and is thought to be the bright core of a galaxy with a massive black hole at its center. The most distant quasar yet detected (designated 0Q172) is thought to be about 18 billion ly (5.5 billion pc) away. Astronomers believe that this distance is very near the outer edge of the universe. If the universe is spherical (and why not?), then 18 billion ly would represent its radius. The known universe would therefore be about 36 billion ly (11 billion pc) across.* What's beyond the universe? Nothing. We've run out of real estate!

How big is 36 billion light-years? Approximately 300,000,000,000,000,000,000,000,000 meters (3×10^{26} m). How small is a proton or a neutron? Approximately .000,000,000,000,1 meter (10^{-13} m). From the smallest (excluding, for the moment, strings and other esoteric particles) to the largest, there are more than thirty-nine orders of magnitude of change. It would take 3,000,000,000,000,000,000,000,000,000,000,000,000,000 (3 thousand trillion trillion trillion) protons or neutrons, placed side by side, like pearls on a string, to stretch across the known universe.

That's some kind of necklace!

---

*This may seem to imply that we are at the center of the universe, but we are not. There is no center, or rather, every place is at the center. This cosmological principle is probably easier to accept than to understand. This concept is touched on in "How Did It All Begin?"

How Dense Is Matter
Inside a Black Hole?

What weighs more, a pound of iron or a pound of feathers? The iron, of course—it is a "heavier" substance. Not so. Since both the iron and the feathers have a weight of one pound, they weigh the same. The difference between the two substances that leads us intuitively to *think* the iron is heavier lies in a physical property of matter called *density*. Iron is not necessarily heavier than feathers, but it is denser. What, then, is density?

Stated simply, *density is how heavy something is for its size*. Expressed more technically, *density = mass/volume*, where mass is the amount of matter that something has, and volume is the amount of space that it takes up. The key to understanding the concept lies with volume. How can feathers and iron have the same weight, or mass? They can if you have more feathers than you do iron. A pillowful of feathers may weigh as much or more than a cupful of iron nails. There are more feathers in the pillow than nails in the cup.

The pillow, therefore, takes up more space. It has more *volume*.

Since volume is important to the concept of density, the best way to compare the density of different substances is to compare the weights or masses of equal volumes. If the volumes are equal, the heavier substance will be the denser. For example, if you had one pillow filled with feathers and another of equal size filled with iron nails, the pillow of nails would be heavier. It would weigh more—it would be denser.

Another way of viewing density is to say that it is a measure of the compactness of matter, or the extent to which matter is squeezed together. All matter consists of tiny particles called atoms, or groups of atoms joined together, called molecules. If these atoms or molecules are close together, the substance is dense. If they are far apart, the substance is not dense.

Let us apply this understanding to the three phases of matter: gases, liquids, and solids. Since gas molecules are spread far apart from one another, gases are not very dense. Scientists would say that they have a very low density. Because there is so much space between gas molecules, they can be easily squeezed together or compressed, increasing the density of the gas. This can be done by increasing the pressure exerted on the gas.

In liquids and solids, however, the atoms and molecules are not far apart; they are practically touching one another. Consequently, liquids and solids are more dense, or have a higher density, than gases. Also, since there is very little space between the molecules, they cannot be compressed very much. In fact, they are practically incompressible— increasing pressure does not increase their density appreciably. (In most cases, solids are slightly denser than liquids of the same substance. Water, however, is a notable exception. Ice is less dense than liquid water. This is the reason why ice cubes float in a glass of water.)

Of the different solids and liquids that exist in nature under ordinary conditions (low pressure and room temperature), osmium is the densest of all. It is a bluish-white metallic solid with a density about 22.6 times that of liquid water. This means that one bucket of osmium would weigh the same as 22.6 buckets of water. Gold is also a dense substance, almost twenty times as dense as water. Mercury is the densest of liquids. In fact, lead will float on mercury because mercury is denser. The average density of Earth itself is about 5.5 times that of liquid water—densest of all the nine planets.

Osmium, then, is the densest substance in nature—*under ordinary conditions*. But conditions are not always ordinary. Under extreme conditions, matter can be squeezed together with great force and can become much denser than osmium. Such densities are "out of this world." Stars are much larger and more massive than planets, giving them a much more powerful gravitational force. This force squeezes together the atoms that make up the star. The more massive the star, the stronger is its gravity, and the more its atoms are squeezed together.

At the center of an atom is a small, dense core called the nucleus. It is composed of two types of particles, protons and neutrons, which are held together very closely— there is virtually no space between the particles in the nucleus. Outside the nucleus is a third type of particle, the electron. There may be from one to more than one hundred electrons in an atom. They circle the nucleus in orbits, or shells, which are spaced far apart from one another (in terms of atomic size). The net result is that most of the mass of the atom is in the small nucleus, with the rest of the atom being mostly empty space. It would be fair to say, in fact, that an atom is mostly (more than 99.9 percent) empty space.

Let us now return to osmium. The atoms of osmium, as of most solids, are practically touching. But it is the electrons in the *outermost shell* that are practically touching. Under ordinary conditions, atoms cannot be compressed beyond this point. The atom is like a foam rubber ball, which cannot be squeezed smaller. This repulsive force of the electrons, resisting further compression, is called the *electromagnetic force*.

In stars, however, the compressing force of gravity may become great enough to overcome the repulsive force of the electrons. For this to happen, a star must be old and cooling. The tremendous heat produced by a young or middle-aged star counteracts the compressing effect of gravity.

Our Sun is a middle-aged, middle-size star. In about 5 billion years it will begin to cool off as it exhausts its nuclear fuel. As this happens, its powerful gravitation will cause the matter that it is made of to compress, or contract. The power of this compression will break through the electromagnetic barrier provided by the electrons. The electron shells will cave in, and electrons will squeeze closer together and begin to move around freely, in a kind of *electronic fluid*. Matter in this state is called *degenerate matter* and is considerably denser than matter in which the atoms are intact. Keep in mind, though, that the electrons in the "fluid" are still spaced fairly far apart. The atoms are still mostly empty space. Degenerate matter can, nonetheless, reach densities greater than one million times that of liquid water. This is about *forty-five thousand times* as dense as the densest substance on Earth. Stars with such density are called white dwarfs. One such white dwarf, known as LP 327-186, is so dense that if a thimbleful of it were brought to Earth, it would weigh more than *one million tons*!

Yet this is not the densest that matter can get. If a dying star is two to three times more massive than the Sun, its gravitational force will be powerful enough to overcome even the repulsive force of the electronic fluid in degenerate matter. The electrons will be squeezed into the nucleus of the atom and will combine with the protons, forming neutrons. Eventually the entire atom will consist of neutrons in contact with one another, in a kind of *neutronic fluid*. Now, for the first time, the atom *is not* mostly empty space. It is a nucleus consisting entirely of neutrons. These nuclei are virtually touching one another, leaving no space for further contraction. Matter in this form is sometimes called *neutronium*.

Stars of this type are called neutron stars, or pulsars. They are very dense indeed—up to one trillion times the density of liquid water, or one million times as dense as the degenerate matter of white dwarfs. In fact, if a baseball-size piece of a neutron star were brought to Earth, it would weigh much more than the Empire State Building. A sugar cube of neutron star material would weigh 100 million tons; if dropped, it would fall through to the center of Earth.

If a dying star has *more* than three times the mass of the Sun, it will have a gravitational force strong enough to crash through even the nuclear barrier of neutronium. There will be no barrier left, no force sufficient to hold back gravity, and matter then will contract until it is a single point, known to physicists as a *singularity*. A star that has contracted to this point is known as a black hole. Who knows how dense matter can be inside a black hole? Astronomers have detected what appear to be black holes in space, and matter outside their center has been estimated to be at least ten times as dense as matter inside a neutron star. The density at the center of a black hole would seem to approach infinity.

Our story is finally over. We have traveled from the

density of intact matter on Earth—a chunk of rock or slice of pizza—to the degenerate matter of white dwarf stars, nearly one million times as dense, to the neutronium of neutron stars, nearly one trillion times as dense, to the ultimate density of black holes, at least ten trillion times as dense. Now that's one heavy slice of pizza!

Can Anything Travel Faster than Light?

Captain James T. Kirk looked at the starboard screen in dismay. "Warp 3, Scotty," he said. "The Klingons are gaining on us."

"Aye, Captain," the first engineer answered, boosting the engines of the U.S.S. *Enterprise* into the *faster-than-light* warp drive.

This "Star Trek"–inspired dialogue may sound familiar to you. In science fiction, faster-than-light travel is accepted as reality. How else could it be possible for humans to journey across the galaxy to explore other stars and other worlds? Traveling *at* the speed of light, the very nearest star (aside from the Sun) is more than four years away, and most stars are *thousands* of years away. But can spaceships travel at "warp" speeds—faster than light? Can anything travel faster than light?

To answer these questions we must first ask, How fast does light travel? This is not an easy question to answer. Light does not always travel at the same speed. Can you run as quickly on sand at a beach as you can on a track? Of course not. Light also has more difficulty traveling through some substances, or media, than it does through others. It travels slower in diamond than it does in glass, slower in glass than it does in water, slower in water than it does in air, and slower in air—though only slightly—than it does in nothing (a *vacuum*). Light, in fact, travels faster in a vacuum than it does anywhere else. But how fast is that?

A beam of light (or any other form of electromagnetic radiation, such as X rays, microwaves, or radio waves) travels at a speed of 186,291 miles/second (299,792 kilometers/second). At this speed, it would travel around the world *more than seven times* in one second! It would travel to the moon in less than two seconds and to the Sun in about eight minutes. In air, light travels nearly *one million* times faster than sound.

The speed of light was first determined with accuracy by a French physicist, Armand H. L. Fizeau, in 1849. The value was refined by a coworker of Fizeau's, Jean B. L. Foucault, and later by Albert A. Michelson, an American physicist who devoted fifty years of his life to the search for the true speed of light. During 1923 and 1924, in California, Michelson performed a classic experiment using a powerful beam of light, mirrors, and two mountaintops 22 miles (about 35 km) apart, measuring with an error of *less than one inch*. He got a value for the speed of light that was accurate to within several thousandths of 1 percent. His work on light helped earn Michelson the Nobel Prize in physics, the first American to achieve this honor.

Let us now return to our original question: Can anything travel faster than light? Since light travels at different speeds in different media, we must restate the question, being more specific: Can anything travel faster than light *in a vacuum*? The answer is no. The speed of light in a vacuum is the limit at which anything—matter or energy—can travel. But why is that? Why can't we take a car or rocket ship and keep on accelerating it, feeding it more and more fuel, making it go faster and faster, until it reaches and surpasses the speed of light?

Take a golf ball, for instance. As it sits on the tee, its speed is zero. When the golfer whacks it with the head of the club, the energy from the moving club is transferred to the ball as energy of motion. A well-hit golf ball may easily attain a speed of 100 miles/hour (160 km/hour). This speed, however, is still about *six million* times slower than the speed of light in a vacuum. But what if the golf ball could be hit six million times harder? Would it not then travel six million times faster, or at the speed of light?

No, it would not. The reasons why this is not possible are complicated and mathematical. Strange things happen as an object approaches light speed. When energy is imparted to an object (such as a golf ball when the golf club hits it), that energy can be used in one of two ways. It can be transferred as energy, causing the object to move faster, or it can be converted into mass, causing the object to have more matter. At all normal speeds, even speeds thousands of times faster than a speeding golf ball, energy imparted to matter is transferred almost entirely as energy. In other words, if you hit a golf ball, it goes faster. As one approaches light speed, however, more and more of that energy is converted into mass, and less is transferred as energy. The object does not speed up as much as it increases its matter. *At* the speed of light, *all* the energy that one puts into an object is converted into mass. A golf ball that is traveling at the speed of light cannot speed up any more. If additional energy were put into it, that energy could not be transferred as energy of motion but would instead be converted entirely into mass. The golf ball would not go faster; it would get more massive. In fact, as one gets very close to the speed of light, so much of the energy is converted to mass that it is virtually impossible for an object to even reach the speed of light in a vacuum.

Scientists have come close, however—though not with golf balls or cars or rocket ships. Using very powerful devices called particle accelerators, scientists have gotten subatomic particles such as electrons, which are extremely small and light and easy to speed up, to travel at speeds close to 186,291 miles/sec.

Another strange thing that happens at near-light speeds has to do with time. As matter travels faster, time slows down. This slowing is not noticeable at ordinary speeds. If you are traveling in a car at 60 mph or in an airplane at 600 mph, time does not slow down enough to be noticed. Your watch would be moving at about the same rate as the watches of people not moving at all. But if you were traveling in the U.S.S. *Enterprise* with Captain Kirk at 186,285 miles per *second*, time would slow down indeed.

Let us say that a friend of yours takes such a journey—travels in a spaceship at a speed of 186,285 miles/sec. You are both twenty years old when she leaves. Fifty years later she returns. You look around for your friend, someone you hope to recognize, even though she is now seventy years old. Instead, you see the same twenty-year-old that you said good-bye to fifty years ago. As she traveled at nearly the speed of light, time slowed down for her and for everyone and everything else on the spaceship. She aged less than a year in the time you aged fifty years. However, she does not realize time slowed down for her. She did not look in a mirror and see herself moving in slow motion. She did not check her pulse and find it beating once instead of sixty times every minute. To her, everything was normal. Everything within her frame of reference slowed down with her, making all seem normal. This phenomenon of time change at increasing speeds is known as time dilation.

We have digressed a bit from the original question— Can anything travel faster than light? We have learned that nothing can travel faster than light *in a vacuum*. But can anything travel faster than light *not* in a vacuum? Curiously, the answer is yes. In 1934 a Russian physicist, Pavel Cerenkov, speeded up electrons to 160,000 miles/sec. in water. Light in water travels at about 140,000 miles/sec. Thus, in water the electrons traveled *faster than light*! For his work, Cerenkov was awarded the Nobel Prize in physics in 1958.

Although light travels enormously fast, especially in a vacuum, it does not travel instantaneously—and this is the source of another fascinating situation. Imagine, for a moment, that you live on a planet in a galaxy so far from Earth that it takes light 100,000,000 years to reach you. (Such distant galaxies *do* exist.) Imagine also that you have a telescope powerful enough to see Earth and what is happening on Earth's surface. Would you see cars speeding on highways, people going to or from work, children at play or in school? No. You would see dinosaurs and other strange prehistoric animals roaming Earth in humid, tropical jungles. This is because it takes time for light to travel. The images that you are seeing through your telescope came from light that left Earth 100,000,000 years ago. These images are 100,000,000 years old.

That light takes time to travel can be evidenced closer to home as well. If you see the Sun rise at exactly 7:00 A.M., it actually rose at 6:52 A.M.—eight minutes earlier. It took light eight minutes to reach your eyes from the Sun. In fact, if the Sun were to suddenly explode and cease to exist, you would still see it shining quietly and steadily for eight minutes.

Light, as you can see, is a truly engrossing subject. From the simple question Can anything travel faster than light? we have converted energy into matter, slowed time, and seen dinosaurs (without having gone to Jurassic Park!). Along the way, we even found time to answer our original question: In media such as glass or water, yes; in a vacuum, no. The speed of light in a vacuum—186,291 miles/sec.—is the fastest that anything in the universe can travel. Sorry, Scotty!

Did Humans Come from Monkeys?

In 1925 a man named John Thomas Scopes was put on trial in Dayton, Tennessee, for teaching the theory of biological evolution. This was in defiance of a state law prohibiting the teaching of doctrines contrary to the Bible. Scopes was found guilty and fined $100.

The trial came to be known as the "monkey trial," because of the widespread belief that evolution meant humans came from monkeys. This, of course, is not true. We did not descend from monkeys any more than we descended from dogs or cats. Rather, at some point in the distant past we all evolved from a common ancestor, branching out like the limbs of a tree, into a variety of new and different organisms. The more recent the common ancestor between humans and another type of animal living today, the more closely related we are. The search for these common ancestors—the study of human origins—is the concern of the science called *anthropology*.

Life began on this planet as a bacterialike microorganism nearly 4 billion years ago. But it wasn't until about 40 million years ago that monkeys first made their appearance on this planet. These early monkeys did not look very much like those of today. In fact, they strongly resembled the lemurlike animals from which they evolved—small, furry things with long tails that lived in trees and foraged for insects by night.

But time and the adaptive forces of evolution wait for no man—or monkey. Soon there were two major branchings from this primitive stock that produced the two varieties of monkeys we see today: New World monkeys, found in Central and South America, are small, have flattened noses, and can swing by their tails, which can wrap around and grasp tree limbs. (Such tails are called *prehensile*.) They probably evolved first. Old World monkeys, of Asia and Africa, are larger, have more prominent noses, and do not have prehensile tails. Old World monkeys also have more advanced hands, with better thumbs for gripping.

It is unquestionably from the line of Old World monkeys that today's humans have evolved. Once again, this does not mean that Old World monkeys gave rise to humans. Rather, we share a common ancestor that lived as recently as perhaps 30 million years ago. (As you read on, refer to Figure 12, the Family Tree of Humans.)

By 15 million B.P. (before the present), the great apes, or *pongids*, had arrived on the scene. Gibbons, orangutans, gorillas, and chimpanzees—today's family of great apes— are our closest living relatives. Of the four, the chimpanzee is our nearest of kin. It is, in fact, more closely related to humans than to most monkeys. DNA analysis shows that 98-99 percent of the chimp's genetic material is identical to our own. Biologists are amazed that with these striking similarities chimpanzees are as different from us as they are.

Because of this genetic likeness, anthropologists believe

Figure 12
Family Tree of Humans
(not drawn to scale)

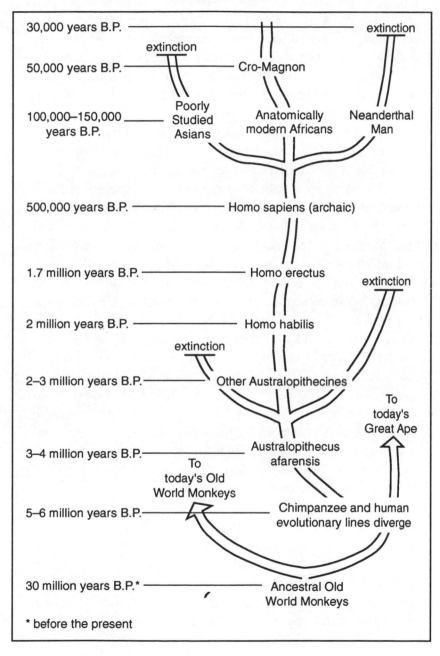

that the split between humans and chimpanzees was fairly recent—about 5-6 million years ago. Since that split, the *hominid*, or human, line has gone through an evolutionary odyssey that has created the human species of today. Along this odyssey, our distant ancestors have left behind bits and pieces of themselves—skeletal remains and teeth—as well as artifacts of their own creation. These fossil bits and pieces have been examined and studied in an attempt to understand where we came from. Although sketchy and incomplete at best, this is the story the bones tell us:

After the great split with the chimps, a species of hominid evolved in South Africa that has been named *Australopithecus afarensis*. *Australopithecus* means "southern ape" and points out the fact that this 3-4 million-year-old hominid was probably more ape than human. In 1974, a remarkably well-preserved female skeleton of *A. afarensis* was unearthed in Ethiopia. Its discoverers named it Lucy because they celebrated her discovery by staying up all night drinking beer and listening to a tape machine blaring the Beatles' "Lucy in the Sky with Diamonds." (Fossil hunters are a strange lot. It is said of one renowned fossil hunter or paleontologist, Robert Broom, that he liked to search for fossils in the nude and was barred from his own research site on several occasions.) Along with Lucy, a dozen other skeletons, including those of children, were found—curiously, all female. The entire group was dubbed the "First Family." Very possibly they were.

Lucy and her kin were about the same size as chimpanzees and had about the same size brain. They did, however, possess one outstanding advance that clearly made them hominids: they were bipedal and walked easily on two feet. Bipedality seems to have been crucial to the evolution of humans. It made it easier for our ancestors to look about for food and avoid enemies. It also permitted males to hunt with weapons and females to carry their young as they moved around. One theory even suggests that the overwhelming right-handedness of humans is the result of females carrying infants with their left arm. This would place the child near the mother's heart, where its beating would have a quieting effect on the youngster.

Over the next 2 million years newer and more advanced models of *Australopithecus* evolved. The skull became rounder and lighter as the brain continued to increase in size. Hands became more modern, with less clawlike nails and a more opposable thumb. *Australopithecus* was, in short, evolving into the genus to which modern man belongs—the genus *Homo*. According to finds in Tanzania (by members of the renowned fossil-hunting Leakey family) and at other East African digs, it appears that the first *Homo* species evolved around 2 million years ago. With a brain approximately half the size of our own and more finely developed hands, he was the first to fashion and use simple stone tools. The importance of this advance in hominid evolution is reflected in the name given to this protohuman, *Homo habilis*—"man, the tool user."

H. habilis hung around for several hundred thousand years, using his tools mainly for cutting wood and butchering carcasses that he found. Not a very efficient hunter, he killed only small game and ate a mostly vegetarian diet. What he might have been best at killing were his *Australopithecine* ancestors, who were still wandering about.

By 1.7 million B.P. or thereabouts, *Homo habilis* had become *Homo erectus*, a taller, brainier version of himself. *H. erectus* was able to create more advanced tools and weapons than *H. habilis*, although they were still quite primitive, even by today's Bushman standards. Sharpened stones were affixed to wooden handles for better leverage, but finer bone implements and the bow and arrow were not yet invented. Nonetheless, these early hominids probably finished off the few remaining *Australopithecines* and by 1 million B.P. were the sole surviving protohumans on the planet.

At about this time, *H. erectus* began expanding his horizons. The fossil record shows that he left South Africa and spread throughout the rest of Africa, Asia, and Europe. When the skeletal remains of the famous Java man and Peking man were discovered, scientists believed that these early human ancestors were the first bipedal hominids. Thus, they were the first to be called *Homo erectus*, or "erect man." Little did we know that our kin had already been walking on two feet for some 3 million years.

When Peking man was discovered, evidence of a fireplace was also unearthed. This was an extraordinary find, for it established *H. erectus* as the first hominid (about 500,000 B.P.) to have mastered fire. The use of fire was a major advance and to this day remains a distinctly human activity. It meant warmth, light, protection from enemies, and the ability to cook—to make previously inedible plant and animal parts edible.

What happened the last half million years, after *H. erectus* started migrating throughout much of the world, is not perfectly clear. We do know that he became more humanlike in appearance—so much so that half-million-year-old fossiis have been found that belong to a new species, the same species to which humans of today belong—*Homo sapiens*, or "wise man."

Shortly thereafter, *Homo erectus* disappeared, leaving to his *Homo sapiens* descendants the task of becoming truly human. They did, but it did not happen overnight. In fact, the name "wise man" hardly seems applicable to these early versions of *Homo sapiens*. For several hundred thousand years they showed very little in the way of innovations or improvements to their way of life. They had no true spoken language, no art, no notable sophistication in technology. Fishhooks, fishnets, boats, sewing needles, musical instruments, and weapons for killing large animals at a distance, for example, are all noticeably absent in the archeological findings of early *H. sapiens*. They might not even have mated face-to-face. How's that for lack of innovation?

Although not terribly creative, archaic *H. sapiens* was adept at distributing himself throughout much of the world. This created populations that, because of their geographic isolation, did not interbreed very much. Consequently, distant groups of *H. sapiens* developed into different subspecies. By 150,000 B.P., *H. sapiens* had evolved into three distinct subspecies. The muscular, brutish, and somewhat apish (although his brain seems to have been 10 percent larger than ours) *H. sapiens neanderthalis*—Neanderthal man—lived in Europe and western Asia. In Africa were people who were increasingly more like us anatomically. And in eastern Asia were people unlike either the Neanderthals or Africans—a group not very well known or studied, since only a few of their bones have been recovered.

What happened next? Several lines of evidence point to an invasion of Asia and Europe by the African group. And the evidence is not just archeological. A new breed of scientists called molecular anthropologists use DNA as a marker to trace human ancestry. In one study, 147 pregnant women from all over the world were persuaded to donate their placentas to science after the birth of their children. Mitochondrial DNA (DNA from the mitochondria and not the nucleus of the cell) was extracted from the placental cells, and their molecular structures were compared. The results were clear: DNA showed the greatest diversity among African women, suggesting that it had accumulated more mutations because it had been around longer. In other words, the African line of descent was the longest branch of the human family tree, implying that from it came modern humans. Because mitochondrial DNA is passed along only by women to their progeny, this hypothetical African mother of humankind was named Eve.

The wanderlust of Eve and her descendants brought them face to face with Neanderthal man and the *Homo-sapiens* of east Asia. Did they wipe out these inferior beings? Was there any interbreeding? Who knows? Perhaps the introduction of unfamiliar diseases did the natives in, much as the American Indians were decimated by smallpox brought to the New World by Europeans.

What we do know (as well as we can know anything from the scattered bones we find) is that sometime about 35,000 years ago—a bit earlier in the Near East—there was what has been called a "great leap forward." Suddenly, where previously there had been Neanderthals in Europe, there were anatomically, fully modern people. They are named *Cro-Magnon* after the French site of their discovery.

Not only was *Cro-Magnon* anatomically our equal, but he was probably every bit as intelligent and innovative as humans of today. Were one transported in a time machine to the present, he or she would be fully as capable as we are of learning to read a newspaper or fly an airplane or work a computer or watch a television soap opera.

The emergence of *Cro-Magnon* brings us to the end of humanity's evolutionary odyssey. But one question still begs to be asked. What abrupt, magical twist in the development of *Homo sapiens* suddenly produced these infinitely more creative beings? What caused that "great leap forward?" Paleontologists theorize that it might have been nothing more than a minor change in the anatomy of their larynxes and throats that allowed them to speak—to create a language with which to communicate ideas. *H. habilis*, almost 2 million years earlier, showed the beginnings of Broca's convolution in the brain—a section governing the power of speech—but neither *habilis* nor his descendants had a throat constructed for talking; that is, not until *Cro-Magnon* came along. A spoken language might have begotten a written language, and a written language would have led to collecting, organizing, and safekeeping of knowledge. The rest, as they say, is history.

Where does humankind go from here? According to a number of noted evolutionary biologists, human evolution has slowed to a halt. "Review the incredible things we've done," says Harvard's Stephen Jay Gould, "how all of civilization has been built in 25,000 years from *Cro-Magnon* to this, with no change in morphology. So why should we predict anything else?" According to Richard Cutler, of the National Institute on Aging, human evolution stopped at least 50,000 years ago. Furthermore, technology has rewritten the survival-of-the-fittest laws. People with defective eyesight who would have been devoured by saber-toothed tigers are now wearing contact lenses and merrily passing along poor vision to their progeny. "Not only are we not improving," says Cutler, "we're getting worse."

Who knows where this nonevolution (if it indeed exists) will lead? In the past, any creature that could not change adaptively perished. Now, in the name of humanity, we are not allowing this to happen. Are we thus preventing our own further evolution? If so, what does it mean? What are the implications? But that's another essay.

Is Earth Getting Warmer?

Have you ever gotten into a car that has been sitting in the Sun on a hot summer day? If you have, no doubt your first reaction was to open the window or reach for the air conditioner button. And with good reason. Although temperatures outside the car may be a comfortable 80° F (27° C), inside it can reach temperatures over 120° F (49° C).

To understand why it gets so hot inside a closed automobile, we must know something about sunlight. The Sun is a vast thermonuclear furnace, spewing out incredible amounts of energy. About one two-billionth of this solar energy reaches Earth's atmosphere, mostly in the form of shortwave radiation such as ultraviolet light, the visible spectrum, and short infrared rays. Roughly half this solar radiation makes it through our atmosphere and strikes Earth.

If your car happens to be sitting in a sunny spot, the

shortwave radiation passes easily through its glass windows, striking surfaces in the car such as the dashboard and the upholstery. These surfaces absorb the radiation and get hot. The heat, however, is trapped in the car, for heat rays are long infrared waves and cannot escape through the glass windows.

A greenhouse, with its all-glass construction, operates in much the same way as the car does. Shortwave radiation is transmitted through the glass, converted to heat rays, and then trapped. Hence, the term for this phenomenon—*the greenhouse effect.*

A greenhouse and a car, however, are not the only places that trap solar heat. Our entire planet is a greenhouse of sorts. Sunlight striking Earth is converted to heat, which is trapped not by a glass enclosure but by certain gases in the atmosphere. The only substantive difference is that whereas the glass enclosure reflects heat back down into the car or greenhouse, the atmospheric gases actually absorb the infrared radiation. This action warms the air blanketing our planet (see Figure 13).

The blanket of air enveloping Earth does more than simply trap heat that would otherwise escape into space. It also distributes the heat more evenly from equator to poles and from sunlit side to dark side. Consider our airless moon, which is roughly the same distance from the Sun as Earth. As one crosses from lunar day into lunar night the temperature rapidly drops from a scorching 212° F (100° C)—the temperature at which water boils—to a frigid −238° F (−114° C).

Chief among Earth's "greenhouse gases" are water vapor and carbon dioxide. Atmospheric water vapor, or humidity, absorbs roughly five times more heat radiating from Earth's surface than do all other gases combined. It is the principal reason our planet is so hospitably warm. Do not,

Figure 13
The greenhouse effect of Earth's atmosphere

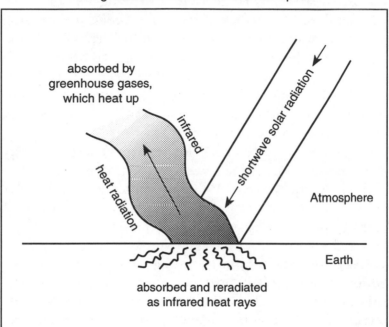

however, discount the other gases, notably carbon dioxide. Carbon dioxide is given off by all living things during respiration. It is also taken in and used by green plants during photosynthesis, a food-making process. Scientists estimate that the mean temperature of our planet would drop from a comfortable 59° F to the chilly mid-40s were it not for carbon dioxide.

For over 100,000 years the two antagonistic processes of respiration and photosynthesis have kept the carbon dioxide in the atmosphere at a fairly constant level of a few hundred parts per million. That is, until about one hundred years ago. . . .

It all started with the Industrial Revolution, nearly two hundred years ago. Huge factories sprang up, each one burning large quantities of fossil fuel (coal, oil, or gas). Like respiration, the burning of fossil fuels releases carbon dioxide into the atmosphere. The buildup had begun.

Then the car was invented. Before long, Henry Ford's assembly line and mass production had created millions of horseless carriages that contributed carbon dioxide–rich exhaust gases to the atmosphere.

The consequence of all this industry and technology was a not-too-surprising increase in levels of atmospheric carbon dioxide. Prior to 1900, carbon dioxide levels never rose above 280 parts per million (based on analysis of carbon from tree rings). The level is presently about 350.

Exacerbating the problem is the fact that carbon dioxide is not the only culprit. Bacteria growing in swamps, rice paddies, and landfills produce another greenhouse gas called *methane*. So do belching cows (over a billion of them worldwide) and termites (three quarters of a ton of termites per person on Earth), which have methane-producing bacteria in their guts. Methane is the principal component of natural gas, which is used widely in homes for heating and cooking. Through the 1980s the level of methane in our atmosphere has risen about 1 percent each year.

Nitrous oxide and the *chlorofluorocarbons* (CFCs) are also being artificially dumped into the atmosphere, and both are efficient absorbers of heat radiation. Nitrous oxide is produced by automobiles when they burn gasoline and by the breakdown of nitrogen-containing fertilizers. "Laughing gas," used by dentists as a pain killer, is nitrous oxide. Chlorofluorocarbons are used as refrigerants in air conditioners and also as organic solvents. Until the banning of their use in the United States and Canada in the late 1970s, they were also the gas propellant in aerosol cans. Although the chief concern with CFCs is their ability to destroy the protective layer of ozone in the upper atmosphere, at lower altitudes they pose significant greenhouse problems.

All told, billions of tons of excess, unwanted greenhouse gases—mainly carbon dioxide—enter the atmosphere each year. Green plants cannot remove, or recycle, this huge overabundance. As a result, Earth is getting warmer.

So what? Aren't winters in much of the world too cold anyway? Wouldn't a warming trend be a welcome change? Not really. Global warming, although it might make winters a bit less brutal in the cooler-climate regions, would spell disaster for much of the world.

Just how severe a warming Earth will experience over the next twenty, thirty, or fifty years, and what the consequences of such a temperature rise would be are impossible to predict. Forecasting tomorrow's weather is hard enough. But sophisticated computer models have been constructed, and they do provide us with some vital information. By the mid-twenty-first century we can expect the greenhouse gases in the atmosphere to be nearly double what they were before the Industrial Revolution. Average temperatures will rise anywhere from 2 to 10 degrees in different areas of the world. One universal consequence of such global warming will be a partial melting of the glaciers and a subsequent three-to-six-foot rise in sea level. Most experts feel that such a rise would be devastating. Low-lying coastal areas such as southern Florida, parts of Texas, and the bayous and wetlands of Louisiana would become submerged. The Caribbean Islands and many other islands throughout the world would pull an Atlantis act and simply disappear. Floodplains would become inundated. Massively populated Bangladesh would lose one sixth of its land area, displacing millions of people. Egyptian settlements along the Nile and the fertile lands of the Mississippi Delta would suffer a similar fate.

Higher sea levels and warmer ocean surfaces mean the

likelihood of more severe hurricanes—perhaps 50 percent more severe. The entire Eastern Seaboard, from Florida to Maine, would be periodically buffeted by high winds and torrential rains that could put New York, Boston, and other major cities under several feet of water.

Another thing the computer models have shown us is that changes brought about by global warming would be far from uniform. The increased temperature would cause greater evaporation of the oceans and a subsequent 5-10 percent increase in precipitation worldwide. But large areas of the planet would actually receive less rainfall—in some cases significantly less. Unfortunately, the United States is one of these areas. Whereas India would receive twice as much rainfall, almost all of North America would become bone-dry. Severe droughts would become commonplace in America as well as major portions of Europe and Asia.

Dramatic shifts in temperature and precipitation, resulting in droughts and floods, would undoubtedly have profound effects on world agriculture. Rice production in the tropics could drop about 10 percent. The warming and drying out of America's wheat belt would significantly reduce our production of that crop, the world's number-one cultivated food supply. Worldwide famine could result. Ironically, two nations that might reap some benefit from this greenhouse heating are Canada and the former Soviet Union—most of their vast lands are at present too cold to be arable. Wouldn't Canadians just love to have America's wheat belt shifted 10 or 15 degrees to the north? Unfortunately, however, things are not that simple: although Canada's climate for crop production might improve, its soil composition would remain just as unsuitable.

Perhaps the most severe upheaval to result from global warming would be ecological in nature. The living things that inhabit any area are in such delicate balance that it is foolhardy to think major climatic changes would not upset that balance irreparably. As Andrew C. Revkin, senior editor of *Discover* magazine, puts it, "Wildlife will suffer too. In much of the world wilderness areas are increasingly hemmed in by development, and when climate shifts, fragile ecosystems won't be able to shift with it. Plants will suddenly be unable to propagate their seeds, and animals will have no place to go. Species in the Arctic, such as caribou, may lose vital migratory routes as ice bridges between islands melt." Who knows what would happen to the tropical rain forests that environmentalists are so concerned about? There would most likely be mass extinctions to rival the disappearing act of the dinosaurs.

Overall, the picture is not an optimistic one. At the Earth Summit in Rio de Janiero in June of 1992, the international community adopted an eight-hundred-page environmental agenda. Unfortunately, however, no limits were set on carbon dioxide emissions. (Chalk one up for industry.) If the greenhouse effect continues unchecked, the planet and all of its inhabitants are in for very hard times. It is hoped that President Bill Clinton's 1993 pledge to reduce America's greenhouse gases to 1990 levels by the year 2000 is not an empty promise.

To totally eliminate the greenhouse problem, our world would have to revert to its preindustrial state. Clearly, this cannot and will not happen. There are, however, a number of measures that can be taken to slow down the warming trend. First and foremost is energy conservation. A 50 percent reduction in the use of fossil fuels would be required to stabilize atmospheric carbon dioxide. Alternative types of fossil fuels must be considered: natural gas, for example, liberates half the amount of carbon dioxide of coal when burned. America, which produces 25 percent of all the world's energy, must lead the way in these efforts, especially in light of the increased levels of energy consumption in emerging Third World nations.

Along with conservation, the more technologically advanced nations must develop solar, wind, and geothermal power. These sources of energy are clean and abundant. Even nuclear power, with all its attendant hazards, might be preferable to fossil-fuel consumption, since it produces no greenhouse gases (only lethal radiation and radioactive wastes).

Much can also be done with automobiles, a major atmospheric polluter. To begin with, we can stop using them so much. In China, one person in 74,000 owns a car; bicycles—300 million of them—are the preferred method of transportation. If we must drive, why not use a car that burns hydrogen? Such a car exists—and you can drink its exhaust fumes, which are pure water. Cars can also run on alcohol, which is easily and cheaply made by fermenting vegetation. Alcohol burning releases two-thirds less carbon dioxide into the air than gasoline does. (One tankful of gasoline generates three to four hundred pounds of carbon dioxide.)

The nations of the world are striving to totally eliminate the use of CFCs by the year 2000 or thereabouts. This is a realistic goal, since CFCs are completely man-made (by a total of thirty-eight companies worldwide) and substitute gases are available.

Deforestation must be restricted. Trees play a major role in removal of carbon dioxide from the atmosphere (photosynthesis—remember?). It is one of the few global processes working for us in our fight against global warming. And burning trees, as with all burning, simply adds carbon dioxide to the atmosphere. Preservation of the tropical rain forest would also save tens of thousands of plant and animal species from extinction.

Trees are not the only photosynthesizers on the planet. Floating in the world's oceans are tiny, green, one-celled organisms that are in such abundance that they account for more photosynthesis than all of the land plants. Were it not for these marine microorganisms, called *phytoplankton*, we would be smothered by a thick blanket of carbon dioxide. But there are certain areas of the world—such as the oceans around Antarctica—where phytoplankton growth is very poor. Wouldn't it be great if we could get these oceans to do their fair share of the atmospheric carbon dioxide removal? Perhaps we can. The research of one oceanographer, Dr. John Martin, has led to the conclusion that algae (a common type of phytoplankton) are not growing in the Antarctic seas because of a lack of iron in the water. His suggestion? Fertilize the ocean with a few hundred thousand tons of iron. The resultant bloom of phytoplankton might go a long way to solving our greenhouse problems.

Dr. Martin's proposal seems a bit extraordinary at first glance. But maybe extraordinary times call for extraordinary measures. We need only to look at our nearest planetary neighbor in space to see what too much atmospheric carbon dioxide can do. Venus, not that much closer to the Sun than we are, has an atmosphere of 90 percent carbon dioxide and a surface temperature that can melt lead. It has, in fact, been called the planet with the runaway greenhouse effect. Although we are in no danger of ever becoming another Venus, our levels of atmospheric carbon dioxide are getting dangerously high.

Are UFOs for Real?

The modern UFO (unidentified flying object) era officially began in 1947. A man named Kenneth Arnold was flying an airplane in the state of Washington when he saw nine disc-shaped objects flying at a fantastic speed. For some reason the newspapers picked up the story and gave it a good deal of press.

Since then there have been well over 100,000 UFO sightings recorded worldwide. In 1987 a Gallup poll revealed that almost half of all Americans believed UFOs were real and one out of eleven claimed to have seen one. If this percentage is accurate, it means that over 25 million people believe they have seen a UFO—an experience referred to as a close encounter of the *first kind*. Clearly, there is something out there. Determining exactly what that something is, however, has proven rather elusive.

The extraterrestrial theory of UFOs is by far the most popular. Scientists have calculated that, based upon the number of suns in the universe and the probable number of

105

planets revolving around them, life almost assuredly exists on many other worlds. So why shouldn't they visit Earth once in a while? Distance, for one thing. Earth is most likely so incredibly far from any planet that might conceivably contain intelligent life that it would take thousands of years for its inhabitants to reach us, even traveling at nearly the speed of light. (Presently, our spaceships can attain speeds of about one ten-thousandth the speed of light.)

But have no fear: If aliens cannot visit us from other worlds, maybe they can be living right on Earth—or in Earth. Another theory of UFOs, popular in the 1950s and 1960s, was the *Hollow Earth* theory. It proposed the existence of a secret civilization that lived inside our hollow planet. Two openings, one at each pole, allowed their flying saucers to enter and exit. By the early 1970s, however, seismographic data and the behavior of artificial satellite orbits showed that Earth was indeed solid and could not sustain a subterranean people.

Whether we conjure up UFOs from outer space or from within Earth or from deep below the sea (yet a third theory), the scientific world is very reluctant to accept such explanations in the absence of hard evidence. (A visitation in which evidence is left behind is known as a close encounter of the *second kind*.) Considering the number of sightings, there is an extreme dearth of such evidence. Photographs are most commonly offered as proof of a visitation. The trouble with photos, however, is their inherent unreliability. A picture of a glowing body or ring of lights in the sky could be anything from a hot-air balloon to a swarm of fireflies to the planet Venus. Modern photographic techniques can also be used to create hoaxes that would deceive the most expert photoanalyst. Suspend a salad plate in the air, provide the proper backdrop and lighting, and you have another close encounter.

Scientists are waiting for some piece of indisputable

physical evidence that an encounter has taken place—a piece of the UFO, the body of an alien. To date, we do not have such proof, although many ufologists are convinced the air force has retrieved and is hiding exactly this sort of thing. What we do have is a ton of circumstantial evidence. Slag and other bits of molten metal allegedly thrown off by UFOs have been collected and studied. Analysis, however, is never conclusive. Although certain unusual alloys of aluminum, magnesium, and iron have turned up, nothing has been found that could not have been manufactured on Earth or deposited by a meteor.

Power outages and the malfunctioning of automobiles and televisions generally show a much higher incidence at the time and place of UFO sightings. Where UFOs have allegedly landed or hovered close to the ground, there are circular scorched or baked areas that cannot satisfactorily be explained as having been caused by lightning or fireballs. High levels of radiation often accompany these scorchings. In one particular alleged contact with extraterrestrials, the space people were health-food enthusiasts. A specimen of the food they cultivated, a potato grown on the moon, was analyzed and found to have five times more protein than potatoes grown here on Earth.

Lunar potatoes notwithstanding, the extraterrestrial origin of UFOs is viewed with great skepticism in the scientific community. What earthly phenomena might be called upon to explain such sightings? Several theories have been put forth, any or all of which can account for the appearance of strange, luminous objects in the sky.

Thermal, or temperature, inversions are one such theory. An inversion is an unusual atmospheric condition in which warm air gets trapped under a cooler air mass. Such inversions can produce optical illusions, or mirages. This might explain the appearance of a UFO but not the interaction that so often accompanies sightings.

On the evening of March 21, 1966, more than eighty women at Hillsdale (Michigan) College saw a glowing object hovering over a swampy area a few hundred yards away. "It was like a squashed football," said one coed. This sighting was attributed to a condition in which rotting vegetation produces the natural gas *methane*. When released suddenly from underground, this "swamp gas" can spontaneously ignite. Unfortunately, most sightings do not occur over swamps, so this explanation can account for only a small number of sightings.

A more likely possibility is that many UFOs are, in reality, ball lightning or glowing plasmas. Ball lightning is a luminous, electrical phenomenon created under atmospheric conditions similar to those that produce normal streak lightning. Glowing plasmas are masses of ionized, or electrically charged, air. They are generated by heavy-duty electrical generators and transmitting equipment and even by aircraft throwing off static electricity. A plasma can exhibit many of the properties associated with UFOs. They make good radar targets, will "follow" cars, often causing them (and televisions) to malfunction, and give off noises as well as UV light capable of producing the skin burns often resulting from too close an encounter.

Whatever theory or theories one subscribes to concerning the existence of UFOs, problems arise when creatures are seen emerging from these spacecraft. And emerge they do. UFO occupants come in every size and shape. Although little green men are a favorite, we also have encountered a Mothman, a Vegetable Man, and "a fire-breathing monster ten feet tall with a bright-green body and a blood-red face." Perhaps the oddest creature was sighted in Lima, Peru, in 1957. A man saw what he described as "two amoebalike creatures that looked like rough-textured bananas" outside a spaceship. They explained to the flabbergasted human that they were sexless and proved the point by promptly dividing in half.

Sightings of living entities are called close encounters of the *third kind*. When the extraterrestrials want to get to know you better, it is a close encounter of the *fourth kind*—contact and interaction. Most CE4s involve abductions of humans by aliens, and they are by far the most fascinating as well as the most terrifying of encounters.

One of the most famous UFO abductions took place in November 1975, in a national forest in Arizona. Seven woodcutters were returning home in the evening when they spotted a brightly glowing UFO in the sky. Suddenly a blue-green beam shot out from the bottom of the craft and struck one of the woodcutters, Travis Walton, and transported him to the spaceship. He was gone for five days. Upon his return to Earth he related a tale of strange-looking humanoids who put him on a table and examined him with an oval-shaped object that was attached to his bare chest. He soon lost consciousness and awoke back on Earth.

Since the Walton incident, many other abductions have been reported. In every case, including Walton's, there is an initial memory block. Only bits and pieces of the abductions are remembered, and there is a dreamlike quality to the whole experience. Obviously, extraterrestrials do not want us to know what they are up to. It is only through hypnotic regression that accurate details of these kidnappings come to light.

But why? What is the purpose of abducting thousands of humans? It is quite evident that the aliens do not want to harm us since very few people—if any—have been willfully killed, and most are returned in reasonably good health. Some, it has been alleged, have even been cured of diseases such as cancer at the hands of the aliens and their strange instruments.

If one is to believe the tales recounted during hypnotic regression, the motive in most abductions is to examine and study humans. Invariably, upon reaching the spaceship, usually through some sort of teletransportation, the abductees find themselves on a table being prodded and poked. The examinations are not particularly benign: Needles are stuck into the brain. Wires are inserted into the penis and the scrotum. Whitley Streiber, a popular novelist *and* abductee, claims that aliens inserted a probe into his rectum and also performed surgical procedures on him. (His experiences are related in detail in his bestseller *Communion*.) Women have had gynecological examinations in which their reproductive organs have been biopsied. Blood, semen, and ova have been taken from various other abductees. In certain instances, these human guinea pigs have thought that implants were placed in their brains, sometimes by instruments inserted into a nostril or an ear. However, no such instruments have ever been recovered.

Sound like medieval torture? Well, in some instances the abductees did report experiencing pain during these procedures. In many cases, however, there was only discomfort. In quite a number of abductions the victims actually felt that the aliens were here to help us in some way. Speaking to abductees telepathically, they showed concern over our destruction of the planet through environmental abuse and nuclear weapons. They wanted somehow to teach us a better way of life. ("Fine, but first get that needle out of my left eyeball.")

A common theme in many abductions is the alien preoccupation with human reproductive systems. This usually meant that aliens took sperm and ova for fertilization and development elsewhere. The literature is replete, however, with acts of sexual intercourse between humans and intergalactic incubi and succubi, the majority of which are humanoid and not always grossly unattractive. In a 1957 abduction in Mexico, a man was seduced by an alien woman who had blood-red underarm and pubic hair. For some reason this aroused him sexually. After the coupling, she pointed to her belly and then the sky, which he took as a reference to their future space child. Kathie Davis, the main abductee in the book *Intruders*, tells of a series of abductions with sexual involvement that took place over a period of years. The first, in December 1977, was for the purpose of inseminating her (artificially—not through intercourse). In March of 1978 she was again abducted and the fetus removed. Finally, five years later, Kathie Davis was transported back to the spaceship and presented with a beautiful daughter, the product of her alien mating. They would not, of course, let her keep the child, whom she described as small for her age, with sparse, patchy hair, and an unusually shaped cranium. She had a triangular face with a tiny mouth and nose and huge blue eyes, strongly resembling her extraterrestrial parent. (See Fig. 14)

Figure 14

Many abductees have taken and passed polygraph tests. Whether their adventures actually occurred or not, the victims sincerely believe they did. And there is some objective evidence to support their beliefs. Many of the abductees have unexplained scars, allegedly the result of operations performed on them. They may also suffer inexplicable radiation poisoning. When kidnappings are reported to have taken place during the night, as they frequently are, there is often blood on bedclothes. In a number of cases there was unexplained healing of old injuries and diseases, mysteriously disappearing pregnancies, and the like. Although other explanations are possible, the question to be asked is, Why would thousands of normal, mentally healthy people go to such extremes to concoct these elaborate fictions? There is no answer—none that makes sense, anyway.

If these visitations are not hoaxes—and the sheer volume of cases suggests they are not—then what the heck is happening? In *Communion*, Whitley Streiber offers several possible hypotheses. Some sound as if they came straight out of the twilight zone. If nothing else, they are fascinating. According to Streiber the visitors could be

1. from another planet or planets.
2. earthlings who have come back from the future. This might explain their interest in helping us.
3. the adult form of the human species. Perhaps death is a time of metamorphosis, in which we, the larvae, transform into mature human beings.
4. beings from another space-time altogether— from a different dimension or universe traveling through what physicists call wormholes.
5. from within us. They could be spiritual projections of ourselves or the collective unconscious

of our species. Many abductees experience the feeling of leaving their physical bodies behind during the abduction. This sort of separation of the spirit from the body (astral projection) is also reported during near-death experiences.

6. elaborate hallucinations. The brain is an electrical device. As such, it produces a faint electromagnetic field in the extra-low-frequency range of 1–30 hertz. Earth itself generates an extra-low-frequency electromagnetic field. Perhaps there are natural phenomena that trigger these wild imaginings in the brain.

One thing is certain: the more we investigate these extraordinary events the less we truly understand. Perhaps the problem is that there has been no serious attempt made by the scientific community to examine the UFO phenomenon. And world governments—the U.S. government in particular—seem to be more concerned with coverups than truthful disclosure. Tabloids make a mockery of the whole thing by purporting to show aliens shaking hands with presidents and presidential candidates. Perhaps the time has come to take this matter seriously.

Can We Win the War Against Cancer?

Cancer is not a new disease. Its markings dot and score Stone Age bones. Tumors are mentioned among the earliest records of Egypt, Greece, and India. In Egypt, between the years 2000 and 1500 B.C., accessible tumors were treated either by excision or application of a corrosive paste. The ancient Romans named the disease *cancer*, the Latin word for "crab," because of the clawlike extensions of a spreading tumor.

Although this dread disease has been around for a long time, we have had relatively little success in dealing with it. Half a million people died of cancer in 1992, nearly ten times more than the number claimed by AIDS. Why has a cancer cure been so elusive?

Simply stated, cancer is the unchecked growth and division of cells. The cells do not grow more rapidly than normal cells. They just do not know when to stop. This uncontrolled cell proliferation produces a cell mass called a tumor.

In time, cancer cells will break off from the tumor and, traveling through the bloodstream, start new tumors growing at distant sites. New tumors seeded from old tumor cells are called *metastases*. Tumors rob healthy cells of nutrients and eventually invade and destroy healthy tissue.

It should be pointed out that there are actually two types of tumors, *malignant* and *benign*. Benign tumors are not cancers. Although they are also produced by continual cell division, they do not invade healthy tissue and do not metastasize. Surgical removal of a benign tumor will effect a complete cure.

Malignant tumors are a different story altogether. Their cells have literally gone mad. Ignoring all biological restraints, they spread throughout the body, destroying everything in their path.

It's in the Genes

Understanding what causes this radical change in a cell's behavior is like learning about life itself. We must study the very atoms and molecules that make up a cell to find out where it went wrong. Every cell has within it a round, dense body called a nucleus. (*Almost* every cell, that is: mature human red blood cells do not.) Contained within each nucleus are the genes—about 100,000 pairs of them. Genes determine everything that a cell will do or become. They are the instructions of life, switching cell functions on or off when necessary. This is accomplished by controlling the proteins a cell makes. Each gene is the blueprint for a specific type of protein. Whenever a cell reproduces by dividing in two, each new cell gets a complete, duplicated set of these vital blueprints. Genes are the hereditary material that pass on the instructions of living from one cell generation to another. A fertilized egg will know how to grow, divide, and develop into a new human being because of the genetic instructions provided by the sperm and the egg cell.

Each gene is actually a minuscule bit of a substance called DNA (short for deoxyribonucleic acid). In the nucleus of the cell, many genes are lined up into long strands of this DNA. Experiments have shown that a cell turns cancerous when there is a change in the structure of the DNA—in other words, when there is a change in the cell's genes.

Any change in the cell's genes, or genetic makeup, is called a *mutation*. Mutations are fairly rare, and even when they do occur, a repair mechanism usually corrects the defect before cell division can pass it on to future cell generations. If the repair mechanism fails, most mutations still will not result in cancer. Only alterations affecting specific genes will turn on the cancer process. These specific genes are called *proto-oncogenes*. When activated by mutation, the proto-oncogene becomes an *oncogene*—a "cancer gene."

Recent studies show that for a healthy cell to become malignant, more than one oncogene must usually be turned on. It is a complex process, which goes something like this (see Fig. 15):

A carcinogen (cancer-causing agent) enters a cell, binds to the DNA, and causes a mutation, which activates an oncogene. Radiation (X rays, for example, or ultraviolet rays from sunlight) can also cause a mutation in the DNA of a cell, activating an oncogene. This first step, called *initiation*, "primes" the cell. Most carcinogens are initiators. Next, a second carcinogen causes the primed cell to proliferate, producing a large population of primed cells. This step, called *promotion*, does not require a change or mutation in a gene. Asbestos fibers are good promoters of cancers initiated by cigarette smoke. This is why cigarette smokers are eight times as likely to get lung cancer if ex-

Figure 15
The making of a cancer cell

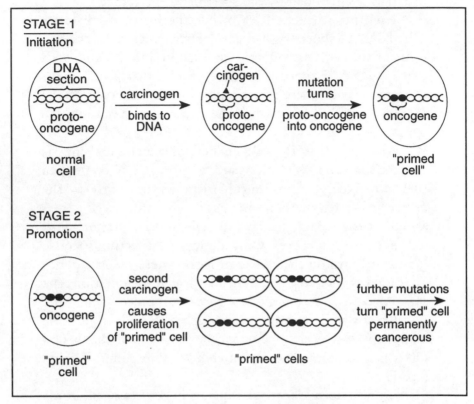

posed to asbestos. Finally, one or more additional mutations to a primed cell occur, turning on yet other oncogenes and permanently transforming it into a cancer cell.

Human oncogenes were first discovered in the early 1980s in a human bladder cancer. Since then, more than fifty oncogenes have been found and implicated in human cancers of the breast, colon, and lung and in certain leukemias. Others are sure to be discovered. There is now little doubt among experts that the switching on of oncogenes causes cancer. The question still being asked is, What is the function of these oncogenes or proto-oncogenes in *normal* cells, and how do they turn a healthy cell into a renegade?

One theory suggests that cancer is a form of cellular regression. During development within the uterus, fetal cells must grow and divide rapidly. Certain genes that promote the production of growth factors prenatally would be actively at work. Upon birth, these genes should become dormant—should shut off. If "switched on" however, later in life, they could become the cancer-producing oncogenes. Support for this theory comes from "fetal proteins," which are released by cancer cells.

Another related theory supposes that the proper function of certain oncogenes is the healing and repairing of wounded tissue. These genes should turn on only when there is tissue damage from burns, lacerations, and the like. If they are incorrectly switched on, cancer results.

In 1993 the gene for a common type of hereditary colorectal cancer was located. Findings indicate the faulty gene causes errors in replication or repair of DNA, leading to nonsense repetitions of tiny, scattered bits of DNA. This may produce mutations in other critical genes, leading eventually to malignancy.

Not only must oncogenes be turned on, but it now appears that certain other genes called *supressors* must be turned off. A supressor gene normally prevents inappropriate cell reproduction. It is a stop signal of sorts. When damaged by mutation, the stop signal is removed and unchecked growth results. Three mutated supressor genes as well as an oncogene have been implicated in one type of colon cancer.

The significance of all of this is that malignant tumors, although all too common, are not easily created; one sudden change in a cell will not do it. There must be several mutations, each necessary yet not sufficient in itself, to cause the cancer. In the case of small-cell lung cancer, as many as fourteen gene changes must take place to cause a malignancy. Oncologists believe this is why cancers take so long to develop, even in people constantly exposed to carcinogens.

Carcinogens and Anticarcinogens

There is a folk ditty that goes like this:

> Everything, everything causes cancer,
> There is no cure, there is no answer.
> Everything, everything causes cancer.

Although this is a bit of an exaggeration, the list of carcinogens is indeed a long one: Pesticides, weed killers, and hormones are but a few. Many carcinogens are found in the foods we eat. A 1984 study listed diet as the leading cancer risk factor in the lives of most Americans. Excess fat in our diet has been linked to breast and colon cancer. Women with family histories of breast cancer are at particular risk. Nitrites and nitrates found in smoked and cured fish and meats promote cancer of the colon and rectum. Burned foods present a significant cancer risk, as do moldy foods. Even supposedly healthful foods such as mushrooms, herbal teas, alfalfa sprouts, beets, celery, lettuce, and spinach contain natural toxins that are powerful cancer-causing agents.

It is believed that carcinogens do their harm by creating highly reactive, electrically charged atoms or molecules in our cells called *free radicals*. During fat metabolism, for example, free radicals are formed. They may react with DNA, causing mutations that trigger the cancer-causing process.

Luckily, however, along with the carcinogens there are cancer-fighting substances in the food we eat. The National

Cancer Institute is investigating hundreds of food chemicals for their anticancer properties. Chief among the anticarcinogens are vitamins A, C, and E and beta-carotene. The mineral selenium is also a potent cancer preventer. They do their work largely by neutralizing the highly dangerous free radicals. Additionally, they bolster the immune system, a mechanism whose importance in combating cancer will be discussed later.

No cancer-prevention diet is complete without a generous helping of cruciferous vegetables and fiber-containing foods. The cruciferous vegetables (so named because their flowers form a cross, or crucifix) include cabbage, broccoli, cauliflower, and Brussels sprouts. They are abundant in potent anticarcinogens, such as indoles and phenolic acids, which somehow block or deactivate a variety of initiators and promotors. Fiber, on the other hand, has no cancer fighting substances. It is, in fact, indigestible, merely providing bulk to speed up elimination of feces. But in its passage through the colon, fiber literally soaks up carcinogenic fats and sweeps them away. Not surprisingly, high-fiber diets greatly reduce the incidence of colon cancer.

Foods are not the only sources of anticarcinogens. According to *The RainForest Book*, "pharmacologists have identified 3,000 plants as having cancer-fighting properties; 70 percent of them grow in the rain forest." Recent research at M.I.T. and Harvard Medical School has shown that shark cartilage can shrink tumors and prevent their spread, or metastasis. It does this by inhibiting *angiogenesis*, the development of new blood vessels in the tumors. Ronald Reagan supposedly was treated for cancer with Carnivora, an extract of the meat-eating Venus's-flytrap plant. Alleged cancer cures have spawned a burgeoning alternative-medicine industry. Unfortunately, the effectiveness of these treatments is as yet unproven.

Certain cancers have been strongly linked to viral infection. Epstein-Barr, hepatitis B, and herpes simplex 2 are three suspect viruses. In fact, the first insights into genetic alteration as a cause of cancer came in the 1970s from analysis of a family of viruses called retroviruses. A number of these retroviruses, known to produce cancers in animals, were found to be carrying cancer-causing genes.

Carcinogens are everywhere, and some have quite interesting origins. In 1775, British physician Percival Potts correctly attributed the high rate of testicular cancer in chimney sweeps to the collection of soot in scrotal skin. Even before Potts's discovery, a Scottish physician linked the use of snuff to cancer of the nose, mouth, and throat. Today, smoking is recognized as the single major cause of cancer mortality, accounting for 30 percent of all deaths. (Diet accounts for 35 percent, but it includes many different foods.) Lung cancer is the leading cancer killer, and 85 percent of all lung cancers are due to smoking. And small wonder: smoking produces a veritable cornucopia of carcinogens that both initiate and promote cancerous mutations as well as suppress the immune system.

The Body Fights Back

The immune system: it is now a household term thanks to the AIDS epidemic. (HIV, the virus that causes AIDS, effectively destroys the immune system.) As far back as 1890, a surgeon named William B. Coley discovered the immune system's amazing capacity for combating cancer. In reading the medical literature of the time, he noticed that spontaneous tumor regression often coincided with an unrelated bacterial infection from which the patient was suffering. Soon Dr. Coley began injecting patients with bacterial toxins. He surmised correctly that the toxins were jolting the immune system into action against not only the invading bacteria but the cancerous cells as well. His cure rates for the time were remarkable but were soon discarded in favor of chemotherapy and radiation therapy.

The workings of the immune system are only now beginning to be appreciated. With the sole exception of the nervous system, it is the most complex system in the body, defending us with a dazzling array of artillery.

According to the *immune surveillance theory*, cancer cells arise constantly in all normal tissues, but the immune system usually recognizes and destroys them. The immune response begins when the body is invaded by a foreign particle or substance, called an *antigen*. Viruses and bacteria are common antigens; so are cancer cells. White blood cells known as *B-cells* somehow recognize the attackers as foreign to the body. Researchers believe that this recognition involves the proteins coating the surface of the intruders. The B-cells, mobilized into action, start churning out *antibodies*—chemicals that bind specifically to the recognized antigen. Antibodies either destroy the invaders outright or tag them so that a variety of killer cells can more easily attack and demolish them. The arsenal of our immune system would be the envy of any country's defense network.

No less important than the cellular component of the immune system are the biologicals—chemical killers that are produced by the immune system. Chief among these biologicals are interferon, interleukin, and the tumor necrosing factor. Interferon, discovered in 1957, has been the most widely investigated because of its strong antiviral properties. At the time of its discovery, a trillionth of a gram of interferon cost $100. Now thanks to mass production through genetic engineering, the cost has dropped dramatically. Much work on these biologicals still needs to be done.

In the movie *Manhattan*, Woody Allen's character says

to his girlfriend, played by Diane Keaton, "I don't get angry, okay? I mean, I have a tendency to internalize—that's one of the problems I have. I—I grow a tumor instead." This line humorously points out what many believe is an important truth: mental health and psychological well-being play a significant role in cancer prevention. The idea is not a new one. The healthy mind–healthy body concept was proposed by the ancient Greeks. But only recently has this idea been put to serious scientific investigation. And the results are clear: stress, depression, and a sense of hopelessness act directly to suppress the immune system and, in turn, one's ability to fight cancer.

What Lies Ahead?

The future of cancer research looks promising. Results, however, will not come overnight and will not come easily. Oncologists today believe that harnessing the cancer-fighting capabilities of our immune system and manipulating the genes that cause malignancies are two directions cancer prevention and treatment will take as we head into the next century.

Dr. Mark I. Greene of the University of Pennsylvania is fashioning an antibody that will bind to and inactivate the protein produced by a certain breast cancer gene. The targetted protein triggers uncontrolled cell growth in tumors.

Researchers have been able to fuse a certain type of cancer cell with an antibody-producing B-cell. The result is a hybrid that acts as a minifactory, churning out huge quantities of an antibody specific for that cancer. These antibodies—called *monoclonal* because they are derived from one line of cells—may someday be used on a wide scale for cancer therapy.

At Scripps Clinic, in La Jolla, California, Dr. Dennis

R. Burton is taking a different tack in efforts to mass-produce human monoclonal antibodies. In a marriage of immunology and genetic engineering, he hopes to isolate the genes that code for cancer-cell-specific antibodies and implant them into bacteria. These bacteria would then become the minifactories for antibody production on a grand scale.

Gene therapy is yet another approach, and perhaps the true wave of the future. It involves altering the genetic makeup of the tumor cells themselves. In one study a herpes virus gene was introduced into brain tumor cells (by a second virus, which is called the *vector*). When enough tumor cells incorporated the herpes viral gene into their own DNA, they became susceptible to the antiherpes drug ganiciclovir, and could be destroyed.

Ultimately gene therapy will be able to correct flaws in the genetic material of malignant cells, replacing cancer-causing genes with normal, unmutated ones. But this will require a much greater understanding of our genetic material (DNA), and how it functions is of critical importance. The Human Genome Project, an ambitious, $3 billion undertaking, is attempting to map every gene on the six-foot coiled strand of DNA found in each cell nucleus. It is a Herculean task, and cancer is a most elusive adversary. In the words of David Baltimore, winner of the 1975 Nobel Prize in physiology, "If I had to choose one field in which the promise of molecular biology has been most evident, I'd have to say it's the study of cancer."

Can Computers Think?

In 1943 the first electronic computer was constructed. Called COLOSSUS, it employed 1,500 electronic tubes, was programmed by two rolls of punched paper tape, and spent its day busily cracking German codes during World War II. Soon other computers with even more vacuum tubes were built. These huge affairs were the first generation of electronic computers. By 1947 the transistor replaced the vacuum tube and ushered in the second generation of computers—smaller, faster, and more powerful. Since then, there have been third-generation computers with integrated circuits and fourth-generation computers with what is termed "very-large-scale integrated circuits."

At each new level the capabilities of the computer increased. Outstanding problems in physics, engineering, and chemistry (such as applying quantum theory to solve the structure of atoms and molecules) fell before the power of the electronic computer. Space exploration, nuclear power, and even accurate weather forecasting would be impossible without computers. But can a computer *think*? Are any of these machines truly intelligent?

That, of course, depends on how we define intelligence. In the early days of computer design, Alan Turing, a pioneer in the field, set a single requirement for machine intelligence. It was simply that, in a teletype communication with a machine, he would be unable to tell that it was not a human being. In 1957, a group of computer experts at Dartmouth College decided that the better a machine was able to mimic human behavior, the "smarter" it was. The outcome of this gathering was a new scientific discipline called *artificial intelligence* (AI). At this meeting, great things were planned for the machines of the future. Just how well have these expectations been realized?

Achievements of Artificial Intelligence

In many respects, AI has had considerable success in creating intelligent machines. We have computers that can give chess grandmasters a good game. In 1988, the greatest chess-playing program, Deep Thought, shared first prize in a tournament, defeating a grandmaster for the first time (though it was defeated the following year). This is no small accomplishment when you consider that in the conventional chess game there are 10^{120} possible move combinations. That's more move combinations than there are atoms in the universe or seconds since the universe began. There are computers that can function as well as any physician in making complex medical diagnoses. When given a test by the *New England Journal of Medicine*, these expert systems performed better than flesh-and-blood doctors. Other expert systems have been created for different fields and are currently being used to prospect for minerals, trace faults in machines, assist in education, and act as travel agents, accountants, and lawyers. Consider the following exchange between a young woman and her computer therapist. The computer has been programmed to recognize key words and transform a client's statements into questions.

WOMAN: My boyfriend made me come here.
COMPUTER: Your boyfriend made you come here?
WOMAN: He noticed that I'm depressed most of the time.
COMPUTER: I am sorry to hear you are depressed.
WOMAN: He's right. I am unhappy.
COMPUTER: Do you think coming here will help you not to be unhappy?
WOMAN: I need some help, that much is clear.
COMPUTER: What would it mean to you if you got some help?
WOMAN: Perhaps I could learn to get along with my mother.
COMPUTER: Tell me more about your family.
WOMAN: My mother takes care of me.
COMPUTER: Who else in your family takes care of you?
WOMAN: My father.
COMPUTER: Your father?
WOMAN: You are like my father in some ways.
COMPUTER: What resemblance do you see?
WOMAN: You are not very aggressive.
COMPUTER: What makes you think I am not very aggressive?
WOMAN: You don't argue with me.
COMPUTER: Why do you think I don't argue with you?
WOMAN: You are afraid of me.
COMPUTER: Does it please you to believe I am afraid of you?
WOMAN: It pleases my father to think I am afraid of him.

COMPUTER: What else comes to mind when you think of your father?

Robots now work on the ocean floor, exploring old shipwrecks, and in outer space, where they are investigating our neighboring planets. Several hundred thousand robots are now being employed in factories, performing such tasks as simple assembly, welding, spray painting, and loading pallets. In 1985 a robotic arm under the control of a computer performed the first surgical procedure on a human brain. There are even computers that are programmed to learn from experience and to use that knowledge to solve practical problems. These learning systems have generated new mathematical concepts, "rediscovered" basic laws in physics, improved their vocabularies, and improved their game of poker by playing against experts.

Where AI Fails

Pretty impressive stuff. Impressive, that is, until we consider that the computer has been completely preprogrammed to make its responses. Impressive until we consider what machines cannot do. They can write simple stories and understand newspaper reports that come in over the wire. Yet when confronted with the biblical quote "The spirit is willing but the flesh is weak," one parsing computer interpreted it as, "The wine is agreeable, but the meat is spoiled." No machine yet created can recognize a specific human face, keep up with the patter of a four-year-old, or move about the world with any degree of facility. Tying a shoelace—something a child does with ease—is well beyond the capabilities of the smartest robot. And it is certainly not for want of trying on the part of AI researchers.

The reason machines have such incredible difficulty

with these "simple" human activities is obvious: They are not simple at all. A much larger chunk of our brain is devoted to vision and muscle coordination than to solving mathematical problems. And we have very little idea how the brain accomplishes these feats. Therein lies the problem. We can teach a computer to play chess because we know the rules and the strategies involved. Therefore, we can create a chess-playing program for the computer. The same holds true for creating computer doctors or travel agents. But we do not understand how the human brain sees or hears. We do not understand how it processes vision and touch data to produce proper muscle response—proper body movement. Without this understanding, we cannot create a program for the computer; we cannot instruct it properly on how to see or hear or move.

So, it appears that intelligence boils down to proper programming—program a computer with all the necessary data and it can perform at a human level. But this is easier said than done. Look at how many years it took just to teach a machine to play one game. Teaching it to survive in an unfamiliar, ever-changing world would be infinitely more difficult. Consider what happened when Dr. Marvin Minsky of MIT hooked a computer to a mechanical arm and a TV camera. He programmed the machine to analyze a stack of children's blocks and then rebuild it after it had been knocked down. The device recognized the block pattern but tried to build the stack from the top down, consistently dropping the blocks in midair. It had never been programmed to know the principle of gravity.

This anecdote points out a very important fact: Knowledge is power. For a computer to exhibit simple "common sense," it must store an incredible amount of data about the world around it. For example, if a young child were told that a giraffe was blowing its nose with its handkerchief, the child would immediately recognize this as silly. A computer would not, unless it had enough knowledge about giraffes, nose blowing, and handkerchiefs programmed into it to logically deduce that this was impossible.

One AI researcher is attempting to create just such a "common sense" computer, instilling it with what every school child knows. Called Cyc (short for encyclopedia), the computer should have 100 million assertions programmed into it by 1994. However, there is one major problem with storing vast quantities of data and rules of logic into a computer's memory. The more you load a computer down, the slower it thinks. Remember, our five-year-old knew immediately that a giraffe cannot blow its nose with a handkerchief. It might take an intelligent machine two hours to figure this out. A robot of the future, loaded with all the data it needs to survive in the outside world, could take several days or weeks to figure it out. This won't be fast enough when the robot has to jump out of the way of an oncoming bus.

But what if computers could be made to work at a much faster pace? The technology is in place to do just that. There are already computers that employ superconducting circuitry. Even more revolutionary computers will use beams of laser light instead of electronic signals to perform their calculations. These innovations will increase computer speed 10,000-fold.

Would these super-fast, super-knowledgeable machines ever be able to "think"? Will a computer programmed to recognize a dog or a cat ever truly understand what a dog or a cat is? Will a computer ever be able to create a symphony or a poem? Will a computer ever become powerful enough to experience emotion? Will it ever be able to make an independent judgment or decision? In the movies *Terminator* and *Terminator 2*, the world of the future is governed and controlled by robots that reach such a level of intelligence that they become "self-aware." Is self-awareness possible, or are such computer accomplishments forever to remain science fiction? Do computers work in ways so fundamentally, so qualitatively different from humans that they are, as one AI expert put it, "forever doomed to remain machines that can at best mimic but never generate true thought?"

Who knows? At this point we can only continue to make computers increasingly more powerful and watch for signs of true intelligence. We can also continue to study the human brain, for it is only through an understanding of this incredible three-and-a-half pound organ that we can begin to unravel the mysteries of the human mind.

The Human Brain

The more that neurobiologists learn about the human brain, the more they are astounded by its complexity. It has been called—without reservation—the most complex structure in the known universe. Consider what happens when you spot a friend across the street: Two billion impulses per second flood into your brain via the optic nerve, going to dozens of different brain centers. Your brain compares the visual image with all remembered images of all the millions of people, dogs, cats, chairs, trees you have seen. Once a match is made, shared experiences with that person flood into your consciousness from stored memories. Signals are sent to your lachrymal glands and a tear is shed. Another processing pathway directs your arm and hand muscles to raise and wave hello. Muscles of the throat, lips, and tongue move in just the proper way, and you shout, "Hi, Susan." And all this happens in about two seconds.

Obviously, the human brain is a marvel of processing efficiency. Its processing unit is made of between 100 and

200 billion nerve cells called *neurons*. The ends of these neurons have many fine branchings, called terminal branches at one end and dendrites at the other end. Some cells, such as the Purkinje brain cells, have as many as 80,000 dendrite and terminal branchings. These end branches allow neurons to make connections with many other neurons. One typical neuron, a pyramidal cell, has up to 100,000 connections to other cells. The result is a vast neural network of a million trillion interconnections and an almost infinite number of possible impulse pathways.

It is through this dazzling microscopic meshwork that the human brain does its work. Each neuron accepts thousands of simultaneous signals and instantly determines whether or not to fire and pass them on to its many interconnected neighbors. In this way, the processing pathways spread out in a hundred million different directions. This is a fundamental difference between the human brain and the computer. A computer is basically one-dimensional: it performs one calculation, which then leads to a second and a third calculation, in a linear fashion. It does not—cannot—perform many different processing operations at once, coordinating and integrating the whole affair as it goes along. But a human brain can—it is, in effect, not one computer but a hundred million computers all working together.

The basic unit of function in the brain, as already mentioned, is the neuron. In a computer, it is the silicone transistor. Simply put, a transistor is a switch that very rapidly shuts off and on, creating pulses of electricity through the circuitry to which it is connected. This creates a sort of Morse code that is the language by which the computer stores and recalls data.

The neuron is a completely different animal. Its signals are electrochemical in nature. This means the electric impulse is generated and propagated by chemicals—called *neurohumors*—that are secreted by the neurons themselves. There are dozens of different neurohumors, and the neurons differ in their degree of responsiveness to each one. They do not merely turn on and off, as transistors do, but demonstrate gradations of sensitivity to the neurohumors that impinge upon them. Once again, there is a complexity of processing in the human brain that greatly belies its effortless operation.

Computer designers today are trying to copy the basic design of the human brain. They are hooking up many computers so that they can work together in what is termed parallel processing. Sandia National Labs, in New Mexico, has developed perhaps the most powerful supercomputer. Called TF-1, it is a massively parallel affair, using 1,024 individual processors. The difficulty is not in linking the computers but in developing the hardware and creating the programs to get them to work together effectively. Some researchers, seeing the futility in trying to create human thought from silicone-based intelligence, are even growing neurons on specially prepared surfaces in an attempt to create a man-made biological processing network.

Of all the differences between artificial and human intelligence, between computers and human brains, the major distinction can be summed up in a word: the human brain is *alive*. It is a dynamic system, able to respond in an infinite variety of ways to changing conditions. For example, the human brain is capable of memorizing the equivalent of five hundred encyclopedias worth of data and of recalling any bit of information instantly. Neurobiologists believe this is accomplished by changing the molecular structure of individual neurons. Precisely how this is done and what the changes are is a complete mystery. But one thing is evident: neurons are constantly altering their structure and their behavior, and no two neurons are exactly alike.

So, can computers or computer controlled robots think? In a word, no—they cannot react creatively to new situations. State-of-the-art AI devices are still little more than sophisticated logic engines, blindly applying the rules that have been built into them. How different this is from human intelligence. As Isaac Asimov put it, "That is the glory of the human brain—it can do things for which we are not yet able to write the rules."

What Is the Most
Venomous Animal?

Dr. Marcos Freiberg and Jerry G. Walls, in their book *The World of Venomous Animals*, called it "without doubt, the most deadly animal to be encountered in the United States and other cool northern countries." What animal could they be talking about? The dreaded rattlesnake? The feared Gila monster? Try the honeybee. Freiberg and Walls explain that the honeybee's sting results in almost as many deaths in the United States as the bites of all venomous snakes combined.

Although this is true, the statistics are very misleading. More people die of bee stings because vastly more people are stung by bees than are bitten by poisonous snakes in the United States. Another thing to consider is that most people who die of bee stings usually encounter a swarm of agitated, aggressive bees. Death results from the stings of *many* honeybees, each injecting its bit of poison. A single bee sting is *rarely* life-threatening.

But which *individual* bite or sting is the deadliest? Which animal makes the most powerful venom? These are not questions that are easily answered. To begin with, the deadliest bite or sting does not necessarily contain the most powerful venom. Deadliness depends upon two factors: the toxicity of the venom and the amount of venom discharged into the victim. The venom of the eastern coral snake, for example, is twice as potent as that of the diamondback rattler, yet the mortality rate of its bite is lower, since it delivers much less venom when it strikes than does the diamondback, one of the largest poisonous snakes.

The coral snake and the rattler are two of America's four poisonous snakes, the other two being the copperhead and the cottonmouth. These snakes are to be feared. They are not, however, the world's deadliest. In Australia, Africa, India, and the Far East there are a number of serpents with far superior killing powers. The krait of China and Burma is one such snake. It has been called "the seven-stepper" because of the belief that after being bitten by the krait you can manage only seven steps before death sets in. This, of course, is an exaggeration, but the many-banded krait does have one of the world's most powerful snake venoms—over twenty times as potent as that of the cobra. Fortunately, it does not deliver anywhere near as much venom as the king cobra, which is capable of injecting 120 times the amount needed to kill a man. This largest of all poisonous snakes (sixteen to eighteen feet in length), with its fearsome hooded head, accounts for more deaths than any other venomous animal worldwide.

Australia is home to many of the world's most unique animals. Its isolation and divergent evolution has produced the kangaroo, the koala bear, and many of the world's most venomous snakes. There are more than eighty-five seriously venomous snakes in Australia, and it is the only continent where poisonous snakes outnumber the nonpoisonous variety. The tiger snake and the taipan are two notable examples. Zoologist and naturalist Roger A. Caras says of the taipan: "It is one of the largest venomous snakes in the world [specimens eleven feet in length have been reported] and delivers one of the most massive doses of venom. Not only is the dosage large, but drop for drop it is one of the most deadly in the world. . . . The recovery rate from the bite of the taipan is low—quite possibly the lowest for all snakes."

Africa also is host to a large number of poisonous snakes, including one that many regard as the most dangerous in the world: the black mamba, known in South Africa as the "shadow of death." Judged the quickest of all snakes, it can travel over the ground at a speed of seven miles an hour. This is faster than many people can run. It strikes like lightning and can deliver a load of venom toxic enough to kill ten men.

Although snakes account for the largest number of deaths due to envenomization, they are not the only venomous animals; in fact, their venoms may not even be the most potent. In the oceans off the coast of Australia is a jellyfish called the *sea wasp*, the most seriously venomous creature in the sea—even more so than the fifty species of highly poisonous sea snakes. Jellyfish are odd-looking animals that resemble translucent balloons floating on the surface of the water. Hanging down into the water from these balloons are the tentacles of the jellyfish. In the sea wasp, these tentacles, which can reach a length of thirty feet, have tens of thousands of stinging cells that contain a potent nerve-acting poison. So powerful is the venom of the sea wasp that when it stings a human, death from asphyxiation or heart failure can occur in less than ten minutes. This ranks the sea wasp as one of the deadliest animals in the world.

The stonefish, found mainly off the coast of eastern

Africa and western Australia, is the deadliest of all fish. Its venom, which primarily attacks muscle tissue, is delivered by the stab of one or more spines on its back. The outstanding feature of the stonefish's venom is the excruciating pain it inflicts. Victims scream and writhe, half-mad with agony. They have been known to plunge the stabbed hand or foot into an open fire and to attempt self-amputation. The pain is apparently unbearable. Death occurs probably as much from shock as anything else.

Of the roughly 3,000 kinds of lizards, only two are venomous. They are the Gila monster, of the southwestern United States, and the Mexican beaded lizard. Unlike snakes, they do not inject venom into a victim with one sudden strike. For envenomization to occur, these creatures have to literally chew the poison into the bite. It is a fairly powerful venom, however, and without medical attention estimates of mortality range as high as 25 percent.

How do insects, spiders, and other bugs fare in this menagerie of seriously venomous critters? Bee and wasp stings, although painful, are rarely lethal, and death usually results from hypersensitivity rather than toxicity of venom. The black widow, the most highly venomous of all spiders, delivers a bite that causes varying degrees of illness yet seldom kills. The same, however, cannot be said for the scorpion, cousin of the spider. Its stinger, located on the tip of its tail, delivers a deadly dose of a powerful venom. Several of the most dangerous species of scorpions are found in Mexico, where nearly 18,000 deaths, mostly of children, were reported for the decade of 1940–1949.

Yet the scorpion is not the deadliest of all bugs. In the early 1970s, studies conducted in Israel led to the discovery of an insect an inch or less in length that has amazing killing powers. The insect, known as the *afrur*, delivers a fairly large dose of an incredibly potent and fast-acting nerve toxin, or neurotoxin, when it stings another animal. As one researcher put it, "A mouse that might survive the bite of a cobra for several minutes will die almost instantly when jabbed by one of these insects." Although no figures of human deaths have been recorded, it is almost certain that the sting of this bug would prove highly lethal.

So, which is the most poisonous animal? the taipan? the sea wasp? the afrur? In terms of potency of toxins, probably none of them. In the forests of Central and South America lives a brightly colored frog less than one inch in length that produces the most powerful animal poison known. It is called the arrow-poison frog. Indians of South America pierce the frog with a sharp stick and then roast it over an open fire. This forces the poison, a toxin that blocks nerve-to-muscle impulses, to the surface of the skin, where it collects in droplets. The natives dip their arrows in this poison, the slightest scratch of which can prove fatal to humans. It is estimated that one arrow-poison frog can treat fifty arrows and that one ounce of its toxin can kill 100,000 people.

Although it produces an incredibly lethal poison, the arrow-poison frog is not considered a venomous animal. It is highly poisonous, in much the same way poisonous mushrooms would be if ingested, but not venomous. It has no teeth, claws, spines, or stingers of any kind to deliver the poison, which has a strictly defensive function: to make the frog decidedly unpleasant for predators to eat. If left to its own devices, the arrow-poison frog is quite harmless and would kill no one.

There is another group of organisms that do not have teeth or claws or spines or stingers. They are not even animals; neither are they plants. They are bacteria, living things too small to be seen with the unaided eye. Yet several of their toxins have the distinction of being the *most powerful poisons known*! Most potent are the toxins causing tetanus and botulism. It is estimated that one tenth of an ounce of either of these toxins would be more than enough to kill everyone in the city of New York. Pretty powerful stuff.

The world is teeming with venomous animals. They account for over 60,000 deaths a year. Every large group of animals—coelenterates, insects, spiders, fish, and reptiles, to name a few—has its venom-producing members. We have examined a select number of the most harmful. Unfortunately, however, it is simply impossible to choose a "most deadly." To begin with, there has never been, nor ever will be, a systematic, scientifically controlled study of every poisonous animal. Without this, we must depend upon reports drifting in from all over the world, many of which are highly unreliable. Even if appropriate studies were conducted, too many variables exist: mode of action of the venom, amount of venom introduced, and sensitivity of the victim to the venom, for example. Suffice it to say that it would be hazardous to one's health to pet any of the animals discussed in this essay.

The Uncertain World
of Quantum Theory

Quantum theory, or quantum mechanics, is the greatest triumph of the twentieth century. Its laws and mathematical equations have allowed us to understand how atoms and molecules operate. Without it we would have no integrated circuitry, no lasers, no computers, no nuclear power, no spaceflight and—worst of all—no television. Molecular biology, which promises to unravel the mysteries of life through its study of DNA, is the child of quantum-theory application.

Why, then, is quantum theory so poorly understood? Part of the reason is that in order to interact with the everyday world, we do not have to understand quantum theory. If you throw a ball up in the air, you know where it is going to land. When you play a game of billiards, you know, or can calculate, exactly what will happen when one ball hits another. Behavior of these objects is governed by the laws of motion and the law of gravity, set down by Sir

Isaac Newton several hundred years ago. And these laws are every bit as valid today.

Laws that govern the behavior of macroscopic objects in our everyday world are referred to as the classical laws of physics. But these classical laws break down when we try to describe behavior of the very small atoms and the even smaller particles of which atoms are made. To deal with these infinitesimal objects we need a new set of rules or, to be more precise, a completely new way of looking at reality.

Hot-Body Radiation

The story of quantum theory begins with the physicist Max Planck. In 1897 Planck was trying to understand why the radiation emitted from a hot body does not conform to expectations. According to classical physics, when a body is heated, vibrations of electrically charged particles (unbeknownst to Planck, they were the newly discovered electrons) within the body produce smooth, continuous waves of radiation. Faster vibrations produce higher-frequency waves with greater energy (see "The Colors of Light" for a discussion of frequency). But—and this is the important thing—the charged particles should vibrate equally well at all speeds, producing radiation of all frequencies in equal abundance. This does not happen. As any blacksmith knows, heating a horseshoe makes it glow red or yellow or white, depending on its temperature. Not only is it emitting frequencies of visible light unequally, but there is virtually no emission of the really high frequency radiation—X rays and gamma rays. Why not?

For several years Planck labored with the classical notion of light, trying to come up with the answer. Finally in 1900, disgusted with his lack of success, he took a daring step. He made the assumption that radiation is not emitted as a continuous wave but rather in discrete packets of energy he called *quanta*. Each packet has a certain amount of energy. The higher the frequency of the wave, the greater the energy in its quanta. At very high frequencies (X rays, gamma rays) the emission of even one quantum requires more energy than is available. Hence, at these high frequencies radiation is negligible.

Thus, the world was introduced to the quantum, and science has not been the same since. Eighteen years after his pioneering work, Planck was awarded the Nobel Prize for his insights.

Photoelectric Effect

In addition to hot-body radiation, there was a second phenomenon that could not be explained by the classical view of light waves. When light strikes certain metals it causes electrons to be ejected from the surface of those metals. This is known as the *photoelectric effect*. The color and brightness of the light affect the ejected electrons in different ways.

Color is a function of the frequency of light. Red has the lowest wave frequency of visible light. As you pass through orange, yellow, green, blue, and violet, the frequency steadily increases. Beyond violet is ultraviolet. The strange thing about the photoelectric effect is that electrons are ejected only if the *frequency* of light is above a certain threshold value. For a given color of sufficient frequency, increasing the brightness increases the number of electrons emitted but not their speed. Increasing frequency, however (going from red to violet), does impart more speed to each ejected electron.

These results are easily explained if we consider the radiation to be quantized—packaged in bundles of energy. A quantum of light of a certain color will always have the same amount of energy. Increasing the brightness of the light increases the number of quanta, but each quantum still has the same amount of energy. Only when the color, or frequency, of light is changed does the energy of the individual quantum change.

Now, using quantized light waves let us consider the photoelectric effect. An electron is ejected from the metal when it is struck by a single quantum of light. Below the threshold frequency the quantum does not possess enough energy to knock out an electron. When the threshold color is reached, an electron will be ejected at a threshold speed. Increasing the brightness of this color releases more quanta, all with equal amounts of this threshold energy. More quanta will strike more electrons, ejecting them, but they will all have the same threshold speed. Turning up the frequency of light, on the other hand, increases the energy of each quantum and therefore the speed of the individual electron it strikes. This is how Albert Einstein interpreted the photoelectric effect in 1905; it was for this work, and not his theories of relativity, that he won a Nobel Prize sixteen years later.

The quantum seemed to explain certain phenomena better than classical views of radiation as a continuous wave. But the greatest breakthrough was yet to come.

Bohr Atom

By 1910, an English scientist named Ernest Rutherford had shown that an atom must have a small, dense nucleus of positive charge with lighter, negatively charged particles revolving around it. This came to be known as the solar-system model because of its similarity to the conventional model of the planets revolving around the Sun. There were, however, numerous problems with this model. Once again, it did not behave in a manner consistent with the laws of classical physics.

First, an electron revolving around the nucleus should be emitting radiation continuously as it speeds along. This does not happen. Second, and perhaps even more important, in radiating this energy it should slow down and very quickly (in a billionth of a second) spiral down into the nucleus. This also does not happen. Why not?

Several men came along to help solve the problem, but it all started with Neils Bohr. In 1911 he joined Rutherford's group to work on the problem of the atom. Bohr had been impressed with Einstein's work on quantized light and soon turned to the quantum to explain the energy difficulties of an electron. He postulated that an electron is in a fixed, stable orbit that might be more aptly called an *energy level*. The electron does not lose energy or slow down as long as it remains in this level. The only time it will move out of this level is when it absorbs a specific quantum of energy. This makes the electron jump to a next-higher energy level. It can then emit the quantum of energy as radiation and fall back to its original energy level. In other words, an electron loses or gains energy only when it changes energy levels, not as a consequence of its revolution around a nucleus.

The good thing about Bohr's quantum model of the atom was that it cleared up the mystery of why elements heated to vaporization emit only certain very specific frequencies of energy. These frequencies, in fact, had been used for years to fingerprint different elements. Simply put, Bohr concluded that the orbiting electrons could absorb only quanta of very specific energies. They could also emit only those same quanta when falling back down to their lower energy levels. The specific quanta determined the colors of light emitted.

Bohr won *his* Nobel Prize in 1922.

Mass Waves

The next advance in quantum theory came in 1923, when Louis de Broglie, a French nobleman and amateur physicist, submitted a doctoral thesis to the Sorbonne. In it he put forth the idea that not only can waves of radiation be quantized and behave as particles, but the opposite is also true: tiny particles such as electrons can and do have wavelike properties. De Broglie quickly made use of his idea to explain a feature of Bohr's atom that had puzzled physicists since its creation. Bohr had shown that electrons move around the nucleus in specific orbits at specific distances from the nucleus. Why had nature chosen those particular orbits and not others? De Broglie asserted that it was because the wavelength of the particular electron wave had to be a perfect fit for that orbit.

De Broglie's matter waves were farfetched to say the least, but they interested Albert Einstein, who in turn sent De Broglie's thesis on them to his friend Erwin Schrödinger. Over the next few years Schrödinger used De Broglie's waves to explain many questions and solve many of the inconsistencies in quantum theory. Both won Nobel Prizes—in the late twenties and early thirties, respectively.

Particle/Wave Duality and the Double-Slit Experiment

Nature is never wrong. If an experiment comes up with unexpected results that astound the experimenter, it is the preconceived notions of the experimenter and not the laws of the universe that are wrong. That is what gave birth to quantum theory. Natural phenomena could best be explained by waves that were actually particles and by particles that were actually waves. But this particle/wave duality of nature has some very weird consequences—so weird, in fact, that it prompted Neils Bohr to say, "Anyone who is not shocked by quantum theory has not understood it."

To begin with, when we say that an electron has wave-like properties, we do not mean it behaves like a person on a roller-coaster ride. This would imply that it is actually a particle moving in a wavelike motion. Not true. We mean that at times it *is* a particle and behaves as such and at other times it changes and becomes a wave. Or, perhaps more correctly, it behaves as *both* a wave and a particle at all times. Call it a "wavicle." This does not appear to make sense, but as John Gribbin says in his book *In Search of Schrödinger's Cat*, "Nobody knows how the quantum world behaves the way it does; all we know is that it does behave the way it does." To better understand what he meant, and just how implausible the quantum world is, we must look at the famous double-slit experiment.

The double-slit experiment was first performed in 1801 by an English physician named Thomas Young. It was done to prove that light is a wave. Light of a specific color (frequency) was sent through two very narrow slits that were spaced close to each other. If light were, indeed, composed of waves, then the waves going through the two slits should interfere with one another, producing alternating dark and light bands on a screen. This type of interference pattern is the signature of a wave. The light in Young's experiment *did* produce an interference pattern—a wave.

But what about electrons? Surely if a beam of electrons were aimed at the slits, each electron must go through only one of the two slits. They would pass through one at a time like a stream of bullets, not like a smooth, spreading-out wave. If they did this, how could they produce an interference pattern like a wave? The experiment was performed and the beam of electrons did indeed produce an interference pattern. De Broglie's electrons were behaving as waves.

But physicists were not satisfied. Somehow, it was assumed, individual electrons from the electron beam were zipping through different holes and in their mysterious wave form were interfering with one another. So the experiment was altered. The electron gun was slowed down so that electrons were shot out one at a time. Each electron was allowed to go through the whole setup before a second was discharged. It might go through one hole or the other, but it was the *only* electron going through at the time—it could not interfere with any other electron or electron wave. Yet, as the number of electrons slowly added up, the typical interference pattern was produced on the screen. Stranger still, if one hole was closed and the electron was forced to go through the other hole—it had no choice of holes—no interference pattern would form. But how can an electron going through one slit know whether or not the other slit is open? And what difference should a second open slit make anyway? This is the central mystery of the quantum world.

The mystery was solved by postulating that an electron has the ability to be in more than one place at a time. It can exist at both slits simultaneously and in its wave form interfere with itself. This was not, however, a solution easily arrived at. The notion that a particle can exist at many points simultaneously certainly seems absurd. Many variations on the double-slit experiment were performed in an attempt to catch the electron going through one hole or another. Although electrons cannot be viewed as such, methods were devised by which electrons (or other subatomic particles) would leave behind evidence of their passage through one of the slits. Surprisingly, the particles always seemed to know when they were being watched. When no one was looking, a particle behaved as a wave, producing an interference pattern. But as soon as you tried to catch it going through the holes, the wave nature disintegrated and you had a particle behaving as a particle, going through only one hole and creating no interference pattern. In other words, the act of observing forced the electron to choose a hole.

How strange, the world of the quantum. Nothing is real unless it is observed. And it is the act of observing that creates the reality. As science writer David Freedman puts it, "The only way to fix a particle in a single location is to observe it. . . . The act of observation not only reveals a particle's condition but actually determines it, forcing it to select just one of the possible states. . . . There is no reality outside observation. . . . As for what goes on between these observations, physicists only shrug and reply, 'Don't ask.' "

Sound like something straight out of the twilight zone? To point out how bizarre this is when applied to the real world of macroscopic objects, Erwin Schrödinger proposed the following thought experiment over fifty years ago: Imagine, said Schrödinger, that you had a cat in a closed box. Along with the cat was a radioactive particle (atom), a geiger counter, a vial of cyanide gas, and a vial-smashing machine. They are all arranged so that if the atom decays, emitting an electron, the geiger counter will detect it, the vial smashing machine will be turned on, and the cyanide will be released. The cat will then breathe the cyanide and die.

Quantum theory states that the radioactive atom is simultaneously in both the decayed and undecayed state— until it is observed. This means that until the box is opened and examined, the vial is both broken and unbroken and the cat is both alive and dead. Obviously, this is nonsense. It is perfectly all right for atom-sized particles to exist in different states at once, but larger objects must obey the classical laws of physics. This is the seeming limitation of quantum theory—it works well only at the submicroscopic level.

Nothing Is Certain

Before closing the book on quantum theory, we must examine one of the theory's most celebrated truths—the *Uncertainty Principle*. It was the brainchild of Werner Heisenberg, who announced it to the world in 1926 and won a Nobel Prize for it in 1932.

The Uncertainty Principle is another one of those quantum-theory oddities that fly in the face of what makes sense in our macroscopic world. It states that one can never be exactly sure of both the position and the velocity of a particle; the more accurately one knows the one, the less accurately one can know the other. In other words, if we know the exact velocity of an electron, we have absolutely no idea where it is. Conversely, if we know the exact position of an electron, we have absolutely no idea what its speed is. Does it sound way-out? Welcome to quantum theory.

When Heisenberg originally proposed this principle, he showed how it was experimentally impossible to ascertain both properties precisely. To determine an electron's position, you would have to observe it by bouncing a quantum of light off it. But the electron might just as well be kicked by a mule. Collision with a quantum would cause it to go flying off in another direction and at a different speed. In other words, the simple act of looking at the electron must necessarily change it so that its measured velocity is uncertain.

Scientists enjoy a challenge. Certainly there must be a way of determining position precisely without disturbing velocity. Many scientists began playing mind games, or "thought experiments," to try to prove Heisenberg's hypothesis wrong. One of the principle players of these games was Albert Einstein. Despite the important role he played in early development of quantum theory, Einstein could not accept that the universe was uncertain. "God does not play dice," he often said, to which Neils Bohr once replied, "Nor is it your business to prescribe to God how He should run the world."

Einstein was wrong. The position/velocity uncertainty does exist—not because we cannot make instruments of sufficient precision but because uncertainty is an inherent property of tiny particles. Actually, according to Heisenberg's equation, uncertainty is a property of all objects, both large and small. The uncertainty, however, increases as the mass of the object decreases. Here is Heisenberg's equation:

$$\text{(uncertainty in position) (uncertainty in velocity)} = \frac{10^{-27}}{\text{mass}}$$

If we assume an equal uncertainty in both position and velocity, the equation simplifies to

$$\text{(uncertainty in position)}^2 = \frac{10^{-27}}{\text{mass}}$$

Using this equation, it has been calculated that the position of an electron can be determined to an exactness of only 1.1 centimeters. This is 11 trillion times larger than the diameter of the electron itself. A baseball's position, on the other hand, can be determined to an exactness of a billionth of a billionth of a centimeter. That quantum of light that would knock the socks off an electron would have virtually no effect on a baseball.

That's the quantum world: Nothing is definitive. Nothing is certain. Nothing is real. As Isaac Asimov said in one of his essays in *Asimov on Physics*, "the only certainty is uncertainty." It doesn't really make sense, but it works. The mathematics derived to explain behavior of matter and energy at the quantum level has been enormously successful in creating the technology of today.

Think you understand it all? Very good. Where would you like us to mail *your* Nobel Prize?

The Most Successful
Animal on Earth

Over one and a quarter million kinds of animals have been identified so far. It is hard to keep an exact count because biologists are finding new ones every day. When we talk of different *kinds* of animals, we actually mean different *species*. Two animals are of the same species if they can mate successfully. Dogs and cats are of different species because their union (if they were so inclined) would not produce viable offspring. A collie and a German shepherd, on the other hand, are of the same species. They can and often do mate, producing litters of mixed breeds. In much the same way, the different races of humans are all of the same species—*Homo sapiens*.

When one studies the animal kingdom, one is struck by its incredible diversity—from microscopic insects and worms to the huge blue whale, a creature that can grow to 100 feet in length and weigh 150 tons, the equivalent of thirty elephants. Even the largest of the dinosaurs, the brachiosaurus, weighed only thirty-five tons.

How did this tremendous diversity within the animal kingdom come about? The answer, in a word, is *evolution*, a process whereby living things change over many, many generations. This change can come about only if two factors exist: there must be variety among the organisms in a given species population, and environmental pressure must be put on the species to favor one variety over another. In Manchester, England, for example, there is an animal called the peppered moth. In the middle of the nineteenth century this moth exhibited a light coloration. Then the Industrial Revolution transformed England—factories began spewing out great clouds of smoke and soot. This pollution darkened the trunks of trees in the Manchester area. Over a forty-to-fifty-year period, the coloration of the moths gradually changed. The light-colored moths were replaced by darker ones.

The explanation is simple: Moths spend much of their time on the trunks of trees. The light-colored moths, more easily discernible against the darker trunks by the birds that feed upon them, were eaten in great numbers. But not all moths were identical in coloration. A small fraction of them were darker, just as people are of different height or weight or coloration. These darker moths, not as easily detected, survived and passed their dark coloration trait to their offspring. Over many generations the lighter-colored moths were replaced by the darker ones.

The gradual adaptation of an organism to better fit its environment has been going on for a very long time. It is a fundamental process of life, one that has worked to a remarkable degree of precision. There is one species of roundworm with such a specific ecological niche that it is found almost exclusively in human appendixes. Another roundworm has never been found anywhere except on the felt mats on which Germans put their beer mugs. Moths can hear ultrasonic sound in precisely the pitch range of the squeaks made by bats on their nightly forays for food.

Such remarkable adaptation has led to an extreme pro-liferation of life forms on this planet. From nineteen speci-mens of a single type of tree in Panama, 1,200 different species of beetles were collected. Each beetle occupied its own little niche in the ecosystem of the tree. This spreading out of a group of organisms to new and different living conditions is called *adaptive radiation*. It is no wonder that over a span of hundreds of millions of years adaptive radia-tion has achieved the degree of diversity we see today.

Yet in this seemingly overwhelming array of different and distinct species there is order. Remember, all present-day organisms evolved from organisms that lived many years ago. The more recent the common ancestor between two different species, the more closely related they are.

Zoologists spend much time studying and trying to arrange animals into groups based on similarities that they hope reflect common ancestry. (Such zoologists are called *taxonomists*.) The first major division among animals is into vertebrates and invertebrates. Vertebrates are animals with backbones and internal, bony skeletons. They include fish, amphibia, reptiles, birds, and mammals. Most of the ani-mals we are familiar with are vertebrates. As a group they have been quite successful at adaptive radiation, especially the mammals.

Mammals are vertebrates that have hair. They are warm-blooded, breast-feed their young, which they bear live, and have a four-chambered heart. The first mammals were tiny, shrewlike creatures that fed primarily upon insects. They appeared on Earth about 70 million years ago. Through adaptive radiation they gave rise to the vast array of mam-mals we see today. Dogs, cats, rats, squirrels, cows, horses, lions, and tigers are all mammals. There are mammals that fly—bats—and mammals that swim the oceans—whales, seals, and dolphins. Some mammals eat fruits and berries. Others graze on grass. Still others hunt and kill other animals. The anteater is a mammal. Baleen whales, a group that includes the huge blue whale, feed on tiny microscopic organisms called plankton, which they filter out of the ocean water.

We need to look no further than human beings to realize the evolutionary success of certain mammal lines. With our superior brain, erect posture, opposable thumb, and ability to speak we not only have adapted to our environment but have conquered it. Great civilizations have been created with ever-improving technologies. In our spare time, we have even produced great works of art, literature, and music.

This does not necessarily mean, however, that mammals are the most successful animals. In fact, they do not even come close—certainly not if the yardstick is adaptive radiation. As diverse and ubiquitous as mammals seem to be, there are only 45,000 different species of mammals worldwide (about a dozen new ones are discovered annually). The entire vertebrate group to which they belong accounts for only 70,000 of all the known kinds of animals. This is barely 5 percent of the animal kingdom. A staggering 95 percent of all animal species identified so far are invertebrates.

Invertebrates are animals without backbones and internal skeletons. Sponges (yes, sponges are animals), jellyfish, clams, snails, starfish, worms, lobsters, spiders, and insects are all invertebrates. Worms are actually divided into three different groups and are among the most successful at branching out into new environments and evolving new species. It is very possible, in fact, that most of the roundworm species inhabiting Earth have not yet been discovered. They are so abundant that a spadeful of garden soil teems with millions of them. But it is as parasites that the worms have achieved their greatest degree of adaptive success.

A parasite is an organism that lives and feeds on or in another organism. When your dog or cat is not feeling well, quite possibly it is suffering from threadlike roundworm parasites. Wherever you find animals, you will find parasitic worms feeding upon them. Most humans are host to any number of flatworms and roundworms. Different species of these worms are found in virtually every part of the body. The liver fluke, a leaf-shaped flatworm, is so highly specialized that it must spend the first stages of its life living in a snail and then a fish before it can invade a human and settle in the liver. The tapeworm—so called because of its flat, ribbonlike appearance—lives in the small intestine of humans and probably every other species of invertebrate. It attaches itself by means of hooks and suckers to the intestine wall, where it can grow to a length of *60 feet*. So highly adapted is this worm to the parasitic way of life that it has completely lost its digestive system and simply soaks up our digested food like a sponge. Each worm has both testes and ovaries to facilitate sexual reproduction.

There are hookworms that burrow through the skin of our feet when we walk barefoot and trichina worms that embed in our muscles. Long, skinny guinea worms live just under our skin and look very much like coiled varicose veins. The African eyeworm (loa worm) is a tiny roundworm that often migrates to the cornea of the eye in humans, where it can cause blindness. More than one hundred bladderworms were found in the brain of a woman who died of convulsions.

Worms are, indeed, a highly successful group of animals. Otherwise, they would not be around today in such abundance. But they are not the most successful of animals. Far more successful than either worms or vertebrates are insects, the true ruling class of Earth. Insect statistics boggle the mind. There are at any given time about a billion billion insects roaming the planet. That's about 200 million bugs for every living human being. Insect species represent over three-quarters of all known animal types. There are about one million different kinds, or species, of insects presently identified. Beetles alone number 300,000—more than all other noninsect species combined. And biologists believe there are an additional *5 to 10 million* insect species, mostly in rain forests, waiting to be discovered.

These are pretty impressive numbers; yet all insects are basically variations on one fundamental body plan. A tough, protective shell, or exoskeleton, covers a body that is divided into three regions—head, thorax, and abdomen. Three pairs of stiff, jointed legs extend from the body and allow for very rapid movement. Spiders are not insects because they have four pairs of legs. Neither are millipedes or centipedes, for the same reason—too many legs. Insects also have a pair of antennae (spiders have none) and usually one or two pairs of wings.

Variations on this fundamental body plan are extraordinary. Insects have achieved unparalleled evolutionary breakthroughs in finding ways to feed, reproduce, and protect themselves. A case in point is the bombardier beetle. It secretes three different chemicals that it then stores in a reservoir chamber located in its abdomen. When it is attacked by a predator, the chemicals are squeezed into a reaction chamber where an enzyme causes them to react explosively. A boiling hot mixture of stinging liquid and vapor shoots out of a turret that can telescope and bend to fire in any direction.

Other insects have evolved to possess ingenious methods of camouflage, looking exactly like sticks or leaves or even thorns of the plants upon which they live. Some non-poisonous insects evolved into lookalikes of poisonous varieties, effectively scaring off would-be attackers.

So adaptable have insects been in meeting the vicissitudes of life that they truly dominate our planet. When conditions become too harsh for humans or other animal types to exist, there will still be the insects. Dr. W. J. Holland, in *The Moth Book*, stated this quite eloquently:

When the moon shall have faded out from the sky, and the sun shall shine at noonday a cherry-red, and the seas shall be frozen over, and the ice cap shall have crept downward to the equator from either pole . . . when all cities shall have long been dead and crumbled into dust, and all life shall be on the very last verge of extinction on this globe; then, on a bit of lichen, growing on the bald rocks beside the eternal snows of Panama, shall be seated a tiny insect, preening its antennae in the glow of the worn-out sun, representing the sole survival of animal life on this our earth—a melancholy "bug."

Perhaps he is right.

A World Without Friction

Television commercials for automotive motor oils advertise their product's ability to reduce the friction of a car engine's moving parts. Friction, it would seem, is a bad thing. But what exactly is friction?

Simply stated, friction is a force that tends to slow down or stop a moving object as it moves against another object. Friction always acts in a direction opposite to that of the movement, thereby retarding forward motion. Push a penny so that it slides freely along a table. Before very long the penny slows to a stop. This is friction at work. The retarding action also makes machines less efficient. As a machine's moving parts rub against one another, more energy must be supplied to overcome the opposing force of friction.

The exact nature of the frictional force between two surfaces is complex and not completely understood. Surfaces that are not very smooth seem to have a harder time sliding over one another than smooth ones. There is greater friction because rough surfaces have tiny projecting irregularities that interlock as one surface slides over another, thereby impeding movement. One would then assume that polishing surfaces reduces friction. This is true—up to a point. Too much polishing, however, will cause friction to increase. Very smooth surfaces, with microscopic irregularities polished away, will lie extremely close to one another. At this exceedingly close proximity, another force comes into play—the molecular attraction between the two surfaces, an attraction that is electrical in nature.

When surfaces rub together, they do more than slow down. The surfaces also heat up. This was very dramatically demonstrated by Count Rumford in the early part of the nineteenth century. He was in charge of boring brass cylinders for the construction of cannon barrels in Bavaria. One day he noticed that a man accidentally touched the brass cylinder being bored and severely burned his hand. *Just how hot is that cannon*, Rumford wondered? Calling for a pail of water, he asked the men to put some of the brass shavings into the water. Soon the water was boiling.

But you do not have to bore cannons to realize that friction generates heat. Your two hands, pressed together and rubbed rapidly back and forth, become warm because of friction. Sliding quickly down a rope or a pole will produce a friction burn. Saw through a thick piece of wood and the saw may become too hot to touch. There is certainly no lack of evidence that friction produces heat.

The heat produced by friction is rarely desirable. If friction is not reduced in a car's engine the pistons will become so hot that the metal will soften, ruining the engine. The hulls of spaceships returning to Earth from outer space must be made of heat-resistant materials to withstand the extremely high temperatures caused by the friction of the ship's outer surface with Earth's atmosphere. Much of the pain you experience when a dentist drills the decay from your tooth is due to the heat produced by friction. The "painless" dental drills use a jet of water to cool down the tooth as it is being worked upon.

Friction also causes surfaces to wear down. Knives become dull because of friction. A major problem encountered in oil drilling is the wearing down of sharp drill bits due to friction with bedrock. These drill bits must be constantly replaced by new, sharp ones.

So who needs friction? It appears to cause all sorts of problems. It's no wonder that great effort and ingenuity have gone into devising methods of reducing its effects. Lubricants are most commonly employed. Motor oil is a commonly used lubricant that reduces friction of the moving engine pistons against the cylinder walls. Graphite, a grayish, powdery solid, is used to ease friction between moving parts in locks so they can open more easily. Graphite consists of flat, smooth crystals that coat the rubbing surfaces, permitting them to glide easily over one another.

Ball bearings, another invention to reduce friction, works on a very simple principle: rolling friction is not as great as spinning friction. A wheel rolling on a set of tiny balls in a track attached to an axle creates much less resistance than a wheel spinning along the axle itself. Sometimes a thin coating of Teflon may be applied to the bearings to further reduce friction. Teflon is the coating on frying pans that makes them stick-free. It also has the property of greatly reducing friction between surfaces that move along one another.

Perhaps the closest we can come to frictionless motion is movement of an object over a layer of gas. If you have ever played air hockey, you know how easily the puck glides over air streaming across the playing surface. The same principle may be employed in a high-speed ultracentrifuge. By suspending its spinning component between the poles of a magnet, its only contact is once again with the air.

Over the years people have gone to great lengths to reduce friction, if not to eliminate it altogether. Friction-free movement, however, is not possible on Earth. Probably the closest thing to true zero friction is in outer space, where a near-vacuum exists and there is no gravity to press the surfaces of objects together. But what if zero friction were possible on Earth? What if we lived in a frictionless world?

On this frictionless world there would be no children running playfully about, no people walking, no cars or buses or bicycles racing around. Why not? Because without friction this type of movement would be impossible. Shoes and tires must grip the ground—get traction—to move forward. That is why it is difficult to walk or to drive a vehicle on ice. The smooth surface of the ice greatly reduces friction. With friction completely absent, tires would spin but the vehicle would not move; feet would slide along the ground, unable to grip, unable to push forward.

It is a good thing that cars could not move in a frictionless world—if they could, there would be no way of stopping them. Braking systems depend upon friction of a brake shoe against a plate or drum to decelerate a vehicle. No friction, no braking.

Friction is such an omnipresent force that its complete absence would have diverse and far-reaching effects. Dentists would be unable to drill decay from teeth; carpenters could not sand and smooth their handiwork. Both activities require the scraping or wearing down of a surface, which involves friction. Gripping tools such as pliers suddenly become useless, as do screws and nails. The simple act of counting out money would be problematic with an inability to grip. It would be impossible to keep dishes from sliding off the table and furniture from sliding all over the floor. The list of minor difficulties and inconveniences would be endless.

On a much grander scale, Earth itself would suffer from the consequences of frictionless motion. One of these consequences involves the rock fragments that constantly bombard our planet from outer space. These fragments, called *meteors*, range in size from microscopic grains to huge boulders weighing thousands of tons. It is estimated that several billion of them enter Earth's atmosphere daily, headed for a collision with our planet. But only a very few of the largest ones actually crash upon Earth's surface; the vast majority burn up in a burst of friction heat as they speed through Earth's atmosphere. Without friction, these meteors would land upon Earth instead of burning up in the atmosphere. This could create a somewhat hazardous situation.

Such conditions do exist on Earth's moon. Due to its small mass and gravity, the moon has no atmosphere. Meteors, unimpeded by friction, crash upon its surface, producing the many craters that cover the lunar surface. Earth would similarly be covered with tens of thousands of craters if meteors were allowed to pelt it without the protective blanket of a friction-producing atmosphere.

The geology of Earth would undergo radical changes in a frictionless world. For one thing, there would be no earthquakes. Earthquakes come about when huge, unjoined sections of Earth's crust (called *plates*), which are pressed firmly against one another, attempt to move. The tremendous friction between the plates, however, does not allow them to slide freely along one another. There is continued pressure to move because the rock under the crust is molten and fluid. Finally, this pressure overcomes the friction force and the plates jolt forward. This jolt is an earthquake. If there were no friction, however, there would be no buildup of pressure and no earthquake would occur.

The forces of wind and running water wear down rock. This is called *erosion*, and it is one of the processes that level mountains and carve out caves and canyons. It is also the process that produces soil. But the wearing down of rock by air and water is a function of the friction between these substances. On a frictionless world, erosion of rock would cease and soilmaking would be greatly retarded. Without soil there could be no life on land—certainly not life as we know it.

Friction may be a nuisance. It wastes fuel. It heats things up. It wears things down. It is the reason you abrade your knee when you fall. Yet without it, there would be no starting, no stopping, no Grand Canyon. All things considered, we might be a lot worse off in a world where friction did not exist.

How Did It All Begin?

"In the beginning God created the heaven and the earth." Sound familiar? It is the opening line of the Book of Genesis, from the Old Testament of the Bible. In one sentence it explains the origin of the universe according to Judeo-Christian belief. Ancient Greeks had their own thoughts on the matter. To quote the poet John Milton, a scholar of Greek mythology:

> First there was Chaos, the vast immeasurable
>   abyss,
> Outrageous as a sea, dark, wasteful, wild.

Everything was a formless confusion. Then, inexplicably, two "children" were born of this confusion—Night and Death. They, in turn, gave rise to Love, which, in its turn, created Light and Day. Earth and Heaven came next, and so on. The details of the creation myths are vague and never well explained, and they are all a bit romantic.

The problem with any of the ancient "theories" of creation, all presented with religious authority, is that they present but do not explain, or, at best, explain but do not offer proof through observation or experimentation. They are not scientifically valid theories of how the universe began. So, what *does* science have to say?

Before starting out, let us paint a picture of the structure of the universe in increasingly broader strokes, beginning with ourselves. We live on Earth, one of nine planets that circle the Sun. The Sun is one of several hundred billion stars that are clumped together in space, forming a galaxy (of which there are about a hundred billion in all). The galaxy we live in is a pinwheel-shaped structure called the Milky Way. Galaxies are clumped together to form galaxy clusters. The Milky Way is part of the galaxy cluster called the Local Group (which contains about thirty galaxies). Galaxy clusters, in turn, are clumped into superclusters, and superclusters into super-superclusters. These large structures look like interlocking soap bubbles, and may stretch 500 million light-years across the sky. *All* of this, the whole ball of wax, is called the universe. In short, the universe is everything, including all the matter, energy, and space that exists. It even includes time.

Astronomers that study the universe on its grandest scale—past, present, and future—are called *cosmologists*; the science that they study is called *cosmology*. How do cosmologists feel the universe began?

Early History

For several hundred years, until the beginning of the twentieth century, the view of the universe was rather conventional. Matter and energy existed as separate entities. There were forces such as gravity and electromagnetism. There was little scientific understanding as to how it all began or how it would all end.

In 1915 Albert Einstein proposed his *general theory of*

relativity. This provided a set of equations that mathematicians and scientists could use to try to explain the workings of the universe. In 1922, Alexander Friedmann, a Soviet mathematician, worked out a solution to Einstein's equations that could describe the conditions of the universe *very*, *very* early in its history. What were these conditions? According to Friedmann, the early universe was incredibly dense—denser even than a black hole (see "How Dense Is Matter Inside a Black Hole?")—incredibly hot, and incredibly small, a kind of cosmic egg or primordial atom.

In 1927 Georges Lamaître, a Belgian priest and astronomer, reasoned that if the universe was much smaller, denser, and hotter in the distant past than it is today, it is likely that an explosion occurred in that distant past, causing the cosmic egg to fly apart. This sudden expansion would increase the size as well as decrease the temperature and density of the universe, ultimately to the conditions we have today. The "sudden explosion" theory of Father Lamaître has been dubbed the *Big Bang* by astronomer Fred Hoyle, and it is basically the theory of the universe we accept today. But is there any hard evidence to support this theory?

Big Bang—Supporting Evidence

If you look up at the sky on a dark, clear night with a pair of binoculars or telescope, you will see stars, of course, and possibly planets. Stars will look like sharp points of light, and planets will appear to be either points or clearly defined disks. You will also see occasional fuzzy-looking objects that early astronomers called nebulae. They did not know what these objects were or how far away they were. In 1924, American astronomer Edwin Hubble determined that they were vastly farther away than either planets or stars. They were, in fact, outside our galaxy. (All the individual stars we see in the sky are *part of* our Milky Way galaxy.) They were, as it turns out, galaxies in their own right. They were so far away that the individual stars that make them up blended into a fuzzy whole.

Hubble studied many of these galaxies and in 1929 discovered something very odd. The light coming from virtually all of the galaxies he studied was *redshifted*. A phenomenon known as the Doppler effect explains the red shift. (See "The Colors of Light.") Put very simply, light-emitting objects, such as lit candles, stars, and galaxies, give off light of a particular color. If the object happens to be moving rather than standing still, the color will appear different. It will appear redder (redshifted) if the object is moving away from the observer and bluer (blueshifted) if it is moving toward the observer. The fact that *all* of Hubble's galaxies, in *all* directions, were redshifted, could mean only one thing: they were all moving away from us. In other words, our universe is expanding.

Astronomers liken this expansion to a loaf of raisin bread baking in an oven. The loaf is the universe, the raisins are galaxies or galaxy clusters, and the dough is the space between. As the bread bakes, the dough rises; the universe expands. The raisins all move farther away from one another.

Logic dictates that if the universe is continuously expanding, even as you read these words, it must have been smaller in the past. Yesterday it was smaller than it is today. A year ago it was smaller yet. Might there not have been a time in the distant past when the universe was an infinitesimally small object (and, by extension, extremely dense and hot as well)? What, then, might have caused this supersmall, superdense, and superhot universe to start expanding? A gigantic explosion.

So, you see, an observable expanding universe implies the validity of both Friedmann's cosmic egg theory and Lamaître's Big Bang theory.

When might this Big Bang have occurred? If we go back in time, the universe gets smaller. How far back must we go to get to the very beginning—when the universe was so small that it could not get any smaller? How far back must we shrink our loaf of raisin bread?

The answer lies in the *rate* at which the universe is expanding. Hubble noticed from his study of galaxies that the farther away a galaxy is, the faster it is moving away from us. A galaxy one light-year away is receding at a speed of nearly 10 miles/sec. (16 km/sec.). A galaxy 10 light-years away is receding 10 times as fast, or at a speed of nearly 100 miles/sec. For each light-year the speed picks up close to 10 miles/sec. This value is called *Hubble's constant*. (The value does slow down a bit with time—tomorrow it will be a bit smaller than it is today. Thus, it is not *truly* a constant; or, to be oxymoronic, it is a changing constant.) Using Hubble's constant and working backward, scientists have calculated that the beginning of the universe—when everything must have been squeezed together into a single point—was 13–20 billion years ago. (We'll use 15 billion as a round figure.)

The Big Bang seems to be a pervasive concept in cosmology. Is there any support for this theory aside from an expanding universe? Yes—it came in 1965, at the Bell Telephone Laboratories in New Jersey. Two Bell employees, American physicists Arno Penzias and Robert Wilson, made a discovery that earned them a share of the Nobel Prize in physics in 1978. They discovered the energy that was left over from the Big Bang. But how?

At the time of their discovery, it had already been well established that the universe is expanding. It was also known that an expanding gas cools. (In fact, this is the basic principle that explains how air conditioners and refrigerators work.) If the universe is expanding, it would therefore follow that it is also cooling. Starting with the very high temperature of the cosmic egg—estimated by cosmologists at about 10^{32} K or °C—and knowing the rate of expansion and cooling of the universe and how long it has been going on—about 15 billion years—scientists were able to predict the temperature of the universe at present. They came up with a value of 3 K (−270° C or −454° F). This is indeed *very* cold. At this temperature, all the gases in our atmosphere would liquefy and freeze.

Penzias and Wilson were able to detect this temperature coming from deep space and in all directions as a microwave radiation. You may remember from reading "The Colors of Light" that temperature relates to the type of radiation with different wavelengths and frequencies. At 20,000 K, an object emits radiation primarily as ultraviolet light. At 6,000 K, the waves get longer, and the color mix is yellow. (This is why our Sun is yellow—its surface temperature is about 6,000 K.) At 3,000 K, the waves are longer yet, in the red-infrared range. At 3 K, the waves have stretched out to one mm and more, the range of microwaves and radio waves. This is the type of radiation that Penzias and Wilson found. They were testing a radio telescope and receiver for Bell Telephone when, quite by accident, they made their discovery. The radiation came in as a kind of static. At first they thought it was from a local source. But the radiation came from all directions, and at all times. And it matched *perfectly* the 3 K radiation predicted by astronomers as the "heat remnant" of the Big Bang. This *cosmic background radiation*, as it is called, has provided such compelling evidence of the Big Bang that the basic theory has been accepted ever since.

Big Bang—A Closer Look

Let's go back 15 million years, to the universe as it is about to begin—to the cosmic egg. It is difficult to imagine the nature of such a thing. The temperature was incredibly high, as already mentioned. The size was incredibly small—on the order of 10^{-50} cm across. This is far smaller than even the smallest subatomic particles. The density, therefore, must have been enormous, since everything that presently exists had to be squeezed into something so small. (Scientists refer to this incomprehensibly small, dense, and hot point as a *singularity*.) Under these very extreme conditions, matter and energy did not exist separately as they do today. In fact, matter and energy as we know them did not yet exist. There were no such things as molecules or atoms or even protons, neutrons, and electrons. The four basic forces—gravitation, electromagnetism, and the strong and weak nuclear forces— were all fused into one force, called the *unified force* or *quantum gravity*. Time and space did not even exist. They began at the moment of the Big Bang—the moment of creation.

About 10^{-35} seconds (some say 10^{-43} seconds) after the Big Bang, the temperature had cooled from 10^{32} K to 10^{27} K. At this cooler temperature, gravity separated from the unified force.

After the universe was about one second old, the temperature had cooled to a chilling 10^{10} (10 billion) K. That's only seven hundred times hotter than the center of the Sun! At this temperature, order began to establish itself from the cosmic chaos that had previously existed. The strong and weak nuclear and electromagnetic forces had already separated, giving us the forces we know today. Cosmologists also believe that by this time *all* the matter in the universe had formed as well. But it existed as subatomic particles only—protons, neutrons, electrons, and a few more exotic beasts such as positrons and neutrinos.

Between the age of about one second and half a million years, subatomic particles joined to form nuclei—about 75 percent hydrogen and 25 percent helium. Over the next half-million years, the temperature cooled to the point where

Figure 16
Big-Bang Theory Time Line

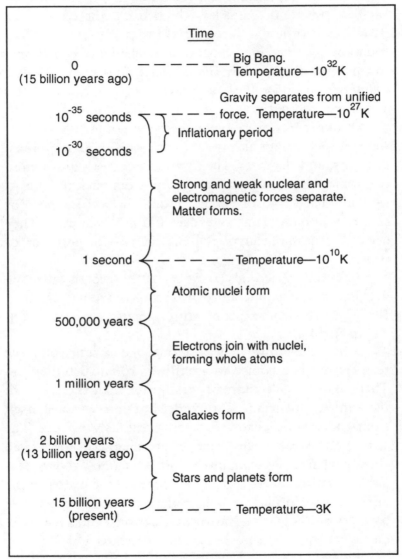

electrons were able to join with nuclei and form whole
atoms. Matter as we know it under "ordinary" conditions
now existed.

The universe continued to expand and cool. Over the next 2 billion years or so, the matter it contained condensed to form huge gas clouds. These clouds concentrated in pockets to form clusters of galaxies.

From about 2 billion years ago to the present, the universe expanded and cooled to 3 K (cosmic background radiation). Stars formed within galaxies, planets formed around some of these stars, and life formed on at least one of these planets. The process is ongoing.

But what about the growth of the universe? We know that it is expanding, and that it has grown from 10^{-50} cm to at least 30 billion light-years in diameter. Did this expansion occur more or less evenly or did it grow in spurts, as we do?

Up until about 1980, the Big Bang theory assumed that the universe expanded more or less evenly through its history. This belief came to be known as the standard model of the Big Bang. Although it fit most observations and calculations about the universe quite well, it failed in a few. (An explanation of these failings is beyond the scope of this essay.)

In 1980 physicist Alan Guth proposed that for a *very, very* brief period of time, when the universe was *very, very* young, it expanded fantastically fast, increasing in size on the order of 10^{30} times. It went from the size of a proton to the size of a grapefruit. (To appreciate this inflation, realize that if a basketball swelled 10^{30} times, it would be 1,000 times larger than the entire universe. This expansion was faster than the speed of light. How can that be? Neither matter nor energy can travel through space faster than light. But matter or energy was not moving through space. Space *itself* was expanding.) After this very brief period of inflation the rate of growth slowed down a lot, and the universe has expanded rather steadily ever since.

Guth's inflationary addition to the Big Bang fits the workings of the universe better than the standard model alone. Cosmologists call it the *inflationary model*, and it is the best thing going regarding the creation and development of the universe. Guth is still awaiting his Nobel Prize.

Cosmology is a fascinating branch of astronomy. As it answers questions, it begs many others. Certain ones can never be answered. How can there be a time when there was no time? If everything that exists came into existence 15 billion years ago, what was there *before* then? If the universe is like a rising loaf of bread, with an outer boundary, what lies *beyond* that boundary? These questions defy conventional logic. They challenge one to comprehend the incomprehensible.

How Will It All End?

Most things have a beginning, exist for a period of time, and then come to an end. The majestic mountain ranges of Earth grow tall from a buckling crust, display their splendor, and then are washed to the sea. The radiant Sun ignites itself from a cold and condensing ball of gases, burns brilliantly, and eventually will flicker out. Life itself is born, exists, and then dies. Is it also the nature of the universe to begin, exist, and end? Certainly it had a beginning. As we learned in the previous essay, the universe was born of chaos and violence, in an unparalleled explosion—the Big Bang. For the 15 billion years since then it has existed. During that time, matter has formed and organized itself into the mosaic of stars, galaxies, and galactic clusters that paint the heavens; energy has straightened itself out, giving us nature's four basic forces—gravity, electromagnetism, and the strong and weak nuclear forces; and the universe as a whole has been getting larger and cooler. But how much longer can this go on? Is the universe subject to constraints that will cause it to end, as mountains and stars end? Or will it continue to expand and cool forever? What is the fate of the universe?

To best understand the fate, or possible fates, of the universe, let us make a simple and concrete analogy. Imagine that a ball is thrown into the air. As it travels upward, it slows down. Eventually it will stop rising and fall back to Earth. The gravitational attraction between Earth and the ball causes this to happen. If the ball is thrown with greater force, it will travel higher before it falls. Can the ball be thrown with sufficient force to overcome Earth's gravity and not fall back down but continue to fly away forever? Yes. If humans could throw a ball with enough force, the ball would not return to Earth. Of course, if Earth were less massive and had less gravity, the ball would require less force, or speed, to overcome the gravity. So whether or not the ball returns to Earth or flies away forever depends on two factors: the speed with which the ball is moving and the strength of the gravitational attraction between Earth and the ball.

Matter within the universe is also flying away. All the galaxies and galaxy clusters are moving away from one another, much as the thrown ball flies away from Earth. The question now becomes, Are the galaxies flying apart with sufficient force so that they will continue to fly apart, or will they slow down, stop, and eventually come back together? As in the case of the thrown ball, the answer depends on the speed with which the galaxies are moving away from one another and the strength of gravitational attraction between the galaxies. Astronomers have accurately calculated the speed of recession of the galaxies (see "How Did It All Begin?"), so whether or not the universe will expand forever depends on the gravitational attraction of the matter within the universe. That, in turn, depends on the amount of matter in the universe. Is there enough of it to cause a sufficient gravitational attraction to stop the expansion and reverse it? Before exploring the answer to this question, let us take a closer look at the possibilities that exist. Keep in mind that although the universe is expanding, its rate of expansion *is* slowing. This is an observed fact.

"Open" Versus "Closed" Universe

If there is not enough matter in the universe, it will expand forever. Cosmologists refer to such a universe as "open." Although the rate of expansion is slowing with time, it will never stop, not even at infinite time.

If there *is* enough matter in the universe, it will stop expanding and start contracting. This contraction would logically continue until the universe is once again an incomprehensibly dense, hot, and small body, as it was prior to the Big Bang. Cosmologists have referred to this squeezing together of the universe back to a mere point, or *singularity*, as the Big Crunch. A universe that operates this way is referred to as "closed," since it reaches a limit in size, with a measurable outer boundary at its greatest expansion. Astronomers have theorized that if there were enough matter in the cosmos for this to happen, the history of the universe may very well be an endless succession of Big Bangs, expansions, contractions, and Big Crunches, with the most recent Big Bang happening 15 billion years ago. This, of course, implies no beginning and no end.

Flat Universe

What if there is enough matter in the universe to halt its expansion but not enough to cause it to contract? The universe would continue to expand and slow, but its rate of slowing would be different from that of an open or closed universe. In an open universe, the expansion would slow but never stop; in a closed universe, the expansion would slow and stop at some time in the measurable future. In this third scenario, the expansion would slow and stop, but at infinity. As with an open universe, this scenario does not allow for a period of contraction or a Big Crunch. Such a universe is said to be "flat." In a practical sense, there is not much difference between an open universe and a flat one.

Now let's get back to an earlier question. Is there enough matter in the universe to close it, or at least to keep it flat rather than open? Scientists at this point do not know, because they do not know how much matter there is in the universe. They do know how much matter there *must* be for it to be flat: 10^{-29} grams per cubic centimeter. (A cubic centimeter is about the size of a sugar cube.) That's one hydrogen atom for every three cubic feet of space. It is known as the *closure density*, since any density greater than this value would cause closure of the universe.

This is not a very dense value, to be sure. In fact, it is a *lower* density than we can achieve on Earth even with our best vacuum pumps, but it is, nonetheless, at least *one hundred times greater* than the density of the universe, taking into account all of the matter that can be seen. The question is, Is there matter out there that *cannot* be seen? If so, where is it, what is it, and is there enough of it to affect the fate of the universe?

Dark Matter

The idea that there is "missing matter" in the universe began in 1933. A Cal Tech astronomer, Fritz Zwicky, had been studying galaxy clusters—in particular, the Coma cluster. He measured the motion of the galaxies in the cluster and came to the conclusion that they were moving too rapidly for the cluster to hold together very long. Yet the cluster remained intact. Zwicky concluded that the cluster must be *ten times more massive* than it appeared in order for its gravity to be sufficient to hold it together this way. In other words, about 90 percent of the mass of the cluster was invisible. Subsequent study of other galaxy clusters yielded similar findings, as did the study of individual stars within galaxies.* Somewhere, somehow, in the seeming void of space, 90 percent of the mass of the universe was hidden from view. Astronomers at first labeled this invisible material "missing mass." But it was not really missing—Zwicky and other astronomers had found it clearly enough. However, it *was* invisible. It did not produce detectable radiation at any wavelength. It was, in effect, *dark matter*. But what is dark matter?

It's easier to say what dark matter isn't. It isn't clouds of hydrogen or helium gas that permeate the cosmos and are the birthplaces of stars and galaxies. These gas clouds would produce detectable radiation at the radio end of the spectrum. It isn't dust either, since dust clouds would obscure light from distant galaxies and therefore make themselves known. It isn't black holes or neutron stars (see "How Dense Is Matter Inside a Black Hole?"), which are strong x-ray emitters. What, then, is dark matter?

Dark matter is probably a lot of things. Brown dwarfs are nearly failed stars with low mass, cool temperatures, and a very faint glow. Though they do radiate energy, it would

---

*Evidence has recently come from another source as well. Intense gravitational fields are able to bend light. Data from certain galaxy clusters indicate that they bend incoming light much more than they should—suggesting a stronger gravity (and greater mass) than their *visible matter* would produce.

not be easily detected. Astronomers calculate that brown dwarfs do contribute to the dark matter in the universe, but not in quantities sufficient to account for Zwicky's mass discrepancy.

Black dwarfs, unlike brown dwarfs, do not radiate energy at all. They are not failed stars but dead stars—stars that have burned themselves out and become cold, dark clumps of matter. Not all stars will become black dwarfs, but most will. The kicker, however, is time: for a star to become a black dwarf would take more time than the age of our galaxy and most others. Hence, it is not likely that black dwarfs contribute in any real way to dark matter.

What about big planets like Jupiter or Saturn? Might not such large planets revolve around other stars? And wouldn't such planets be invisible? After all, planets do not give off their own light. Astronomers agree that there are probably many stars in our galaxy and others that have planets revolving around them, but endless searching of the cosmos with today's highest-tech instruments has so far revealed only *one* verified extrasolar planetary system. This strongly suggests that their numbers are fewer than was once thought and that they do not constitute a significant amount of dark matter.

So where's the beef?

The beef may be not in large bodies of matter, such as feeble stars, dead stars, and planets, but in subatomic particles. High-energy particle physics, a relatively new field of science, is turning up a menagerie of exotic particles that are definite dark-matter candidates. Several of these are photinos, gravinos, gluinos, neutrinos, and weakly interacting massive particles, or WIMPs. Neutrinos have recently fallen into disfavor because of their nearly massless nature, and WIMPs are too difficult to detect to be taken very seriously at this point. Photinos, gravinos, and gluinos have yet to be detected; they exist only theoretically, being predicted in the computations and analyses of high-energy particle physicists.

Enough of dark matter; let's get back to the "matter" at hand. Where do we stand with respect to the universe and how it will end? Presently, astronomers can see 1 percent or less and can detect (through gravitational effects) about 10 percent of the matter that must exist to close the universe. Is there enough *undetected* dark matter out there to account for the difference? If push came to shove, most astronomers would say no. Which is to say that the universe is open and will continue to expand and cool forever. This, of course, implies that the Big Bang was a one-time event.

Recent findings, however, with respect to the 3 K cosmic background radiation, strongly confirm the inflationary model of the Big Bang theory. (Both the 3 K radiation and the inflationary model are discussed fully in "How Did It All Begin?") This model, in turn, predicts that the mass in the universe is *exactly the amount needed for closure*. Which is to say that the universe is flat—it will continue to fly apart, halting at infinity. This also implies a one-shot Big Bang.

Although a closed universe does not seem very likely, it has a sort of symmetry that is appealing: a universe that expands and contracts every 40 or 50 billion years, Banging and Crunching indefinitely, with each fiery end constituting a new beginning.

Are We Alone?

A radio station is playing dance music when the announcer cuts in with a wire report: "9:15 P.M. Eastern Standard Time. Seismograph-registered shock of almost earthquake intensity occurred within a radius of twenty miles of Princeton." It is thought to be from a meteorite impact. But a strange large, metal cylinder lies half-buried in a pit at the impact site. As hundreds watch curiously, the top of the cylinder slowly opens and something tentacled crawls out, all wet and slithery. Several onlookers move closer. On the radio station, a reporter shouts, "Wait! Something's happening." A pause, then a hissing sound. "A humped shape is rising out of the pit. I can make out a small beam of light . . . a jet of flame. . . . It leaps at the advancing men. It strikes them head-on! Good Lord, they're turning into flame!"

Six million people listened to Orson Welles's 1938 radio broadcast of H. G. Wells's *The War of the Worlds*. And more than a million of them panicked and actually believed that Earth had been invaded by Martians. It is now more than fifty-four years later, and we are reasonably certain that there is no life on Mars or, for that matter, on any other planet in the Solar System—no intelligent and technologically advanced life that could travel to us in space ships. But how about life outside the Solar System, on planets circling other stars? Is anybody out *there*?

A Difference of Opinion

The answer depends on whom you ask. Some astronomers feel that life, especially intelligent life, is unlikely anywhere else in the universe. They reason that too many circumstances must be just so, and too many events must follow in just the right order. Others—science writers Carl Sagan and Isaac Asimov included—feel that the creation of life is not nearly so unique and self-limiting and that if intelligent life exists here, it must also exist elsewhere. They reason that with better than 40 billion trillion (40,000,000,000,000,000,000,000) stars in the universe, the Sun (and Earth) cannot be *that* unique. The majority of scientists agree with this view. In 1982 a blue-ribbon panel of astronomers concluded that "Intelligent organisms are as much a part of the universe as stars and galaxies."

In an attempt to quantize the likelihood of extraterrestrial intelligence (ETI), or, more precisely, ETI that can make its presence known to us, University of California astrophysicist Frank Drake (a pioneer in the field) developed the following equation:

$$N = N_* \times f_p \times n_e \times f_l \times f_i \times f_c \times f_L,$$

where

N = number of communicating ETIs in our galaxy, the Milky Way

N_* = number of stars in the Milky Way

f_p = fraction of stars that have planets

n_e = average number of planets circling each star that can support life

f_l = fraction of such planets on which life has actually originated

f_i = fraction of those planets on which intelligent life has evolved

f_c = fraction of those planets on which the intelligent beings have developed the means and desire for interstellar communication

f_L = fraction of a planet's life occupied by such communicating civilizations

Unfortunately, several of the factors in the equation are little better than guesses, which leads to a wide margin for error. Drake himself predicted 4,000 worlds with communicative civilizations. Isaac Asimov, in his compelling work *Extraterrestrial Civilizations*, calculated 530,000.

Let us not quibble with numbers but accept the premise that there is intelligent and technologically advanced life out there. How will we be made aware of their presence? On this point, there is general agreement among astronomers.

No Landing Strips for Little Green Men

Alien life will not make its presence known by landing in some field or backyard. There will be no "little green men" with ray guns crawling out of flying saucers, bent on destruction or looking to phone home. Instead, they will announce their presence through radio communication. It is the only logical way. Using ourselves as examples, we can see why. With present technology, our fastest spaceships can travel at about *one six-thousandth* the speed of light. (Sorry, Trekkies—no warp drive yet!) The nearest star to us aside from the Sun, Proxima Centauri, is 4.2 light-years away.

This means that it would take more than *25,000 years* for a space ship leaving Earth and traveling at maximum speed to reach Proxima Centauri. And most stars are *much* farther away than that. Clearly, we are nowhere near the point in space technology where we can contemplate interstellar travel.

Radio communication, however, is another story. Radio waves travel *at* the speed of light, not six thousand times slower. They do not require expensive space vehicles to launch them—just simple and inexpensive transmitters. And radio technology is not futuristic; it has been around for quite some time. In fact, for nearly three-quarters of a century we have been bombarding space with radio waves. At this moment, intelligent life on a planet circling Proxima Centauri could be tuning into an afternoon broadcast of "Another World" while life *thirty* light-years away could be twitching their hairy little antennae over an episode of "Gilligan's Island" or "The Honeymooners." The ever-expanding shell of our radio transmissions extends outward for more than seventy light-years in all directions, covering many thousands of star systems in our galaxy.

We can only assume that extraterrestrial civilizations have evolved technologically in a similar fashion and that they will be prepared to communicate with us through radio transmissions centuries before they will through vehicular travel. Thus, instead of building landing strips for alien spacecraft, we should be building radio receivers and pointing them to the stars. Are we?

Search for Extraterrestrial Intelligence

In a 1959 article in the science journal *Nature*, physicists Giuseppe Cocconi and Philip Morrison suggested that the best way to communicate across the vast distances of space would be through radio waves—in particular, microwaves.

Unlike visible light and certain other parts of the electromagnetic spectrum, microwaves pass unabsorbed through
the dust and gas clouds that pepper deep space. A few
thousand watts of power can send a radio message clear
across our galaxy, which spans 100,000 light-years.

Radio astronomer Frank Drake was intrigued by the
possibilities of radio communication. The first large radio
telescopes (receivers) had already been built by the mid-
1950s. They were huge concave structures of metallic grid-
work, similar in design to backyard TV satellite dishes. In
1960, Drake pointed one of them, an eighty-five-foot-di-
ameter dish, at two nearby sunlike stars. Thus, the beginning of an organized and concerted effort to detect radio-
communicating extraterrestrial life in our galaxy had begun.
The project was called the Search for Extraterrestrial Intelligence, or SETI, and was under the auspices of the National
Aeronautics and Space Administration (NASA). Over the
past thirty-two years since its inception, SETI has had a
rather uncertain history. Money was not always forthcoming, and many valuable aspects of the program had to be
scrapped or put on hold. Nonetheless, it limped along,
panning the skies for radio signals from intelligent life on
other worlds.

Astronomers in the SETI program tuned their radio
receivers to different frequencies, or channels, much as you
might change channels on a TV or a radio receiver at home.
The radio part of the spectrum, however, is rather large, and
astronomers had to decide which frequencies were most
promising. Two of SETI's favorites were

- microwaves with a frequency of 1420 megahertz
  (1420 million oscillations per second) and a
  wavelength of 21 centimeters
- microwaves with a frequency of 1720 megahertz
  (1720 million oscillations per second) and a
  wavelength of 18 centimeters

The region between and including these frequencies is called the Water Hole. A water molecule (HOH) consists of a hydrogen atom (H) and a hydroxyl group (OH). When hydrogen atoms (actually, H_2 molecules) collide in space, they radiate energy; this energy has a frequency of 1420 megahertz (MHz). When OH groups collide in space, they radiate energy with a frequency of 1720 megahertz (MHz). Hence the term Water Hole.

But why would aliens prefer these wavelengths for their broadcasts? First of all, hydrogen is the most abundant element in the universe. And water is necessary for life as we know it. We know these things. Intelligent aliens would know them as well. Radio emissions associated with hydrogen or the components of water molecules would provide a common ground, a familiar place to operate at. Though this line of reasoning may not mean anything, it makes sense. Certainly, it makes more sense than randomly choosing a frequency.

Secondly, the Water Hole is a radio-quiet part of the spectrum. Despite the abundance of hydrogen in the universe, the radiation that it emits is rather weak. Other portions of the radio band, on the other hand, are plagued with more intense radiation—background "noise" from various sources that would interfere with signals from radio-smart aliens. The galaxy produces its own brand of radiation from electrons being accelerated in magnetic fields. Then there is radiation left over from the creation of the universe. (See "How Did It All Begin?") Radiation is also generated within Earth's atmosphere. Yet none of these radio frequencies are within the Water Hole. It is quiet there—quiet enough to hear faint whispers.

So why haven't we heard from at least *one* bug-eyed monster or intelligent blob? Perhaps we haven't been listening closely enough. But that has all changed now. SETI is no longer on the back burner of NASA's agenda.

Microwave Observing Project

At precisely 3:00 P.M. Eastern Standard Time, on October 12, 1992, switches were flipped at Arecibo, Puerto Rico, and Goldstone, California, turning on two of the most powerful radio telescopes in the world. It marked the beginning of a renewed and entirely revitalized SETI program, the Microwave Observing Project (MOP). Not entirely by coincidence, the date also marked the 500th anniversary of Columbus's landing in the New World. As Columbus searched for a new world, so the scientists of Earth are searching for life on new worlds.

Commitment to the project is genuine and well funded. The search will span ten years and cost $100 million. The technology being used is the very best, including the largest radio telescopes hooked up to the most sophisticated computers. Whereas SETI of the past tuned into several frequencies at a time, MOP will tune into *more than ten million channels* at a time. The information-gathering potential of the project is staggering. MOP can acquire more data in thirty seconds than all the data that has been acquired in the past thirty years.

Also, the project's two-pronged search strategy is much more comprehensive than earlier SETI programs, so as to leave no stone unturned. The first prong, called the Targeted Search, employs three huge radio telescopes—the world's largest (the dish at Arecibo, which is more than three football fields across) and two that are somewhat smaller. Working under the assumption that sunlike stars are more likely to have planets around them with radio-intelligent life (after all, *we're* here), astronomers have pointed these telescopes at solar-type stars. In the ten-year lifetime of MOP, they expect to log close to 1,000 such stars, or just about all that there are within a range of 75 light-years. The channels they are tuning in to are in the region of the Water Hole. Astronomers feel that the Targeted Search is their best shot at finding the alien needle in the cosmic haystack.

But what if our assumptions are wrong? What if radio-smart life is just as likely, or more so, around *non*sunlike stars? After all, stars like our Sun comprise less than 10 percent of all stars. And what if radio-intelligent aliens choose to broadcast *outside* the Water Hole? What if the Water Hole is all wet? The reasons for choosing Water Hole frequencies in the first place are tenuous at best. Hence, the second prong of the MOP search strategy. It is called the Sky Survey, and it is more broadly based in its approach. Over the ten-year life of MOP, it will search everywhere and tune in to frequencies from 1,000 to 10,000 MHz—a much wider region of the radio spectrum than the Water Hole. The workhorses of the Sky Survey will be the 230-foot (70-meter) radio telescope at Goldstone, California, and another smaller scope at Tidbinbilla, Australia.

These are the technological aspects of the Microwave Observing Project. But there are philosophical aspects as well. So what if there's intelligent life out there? Who cares? With millions of people unemployed or homeless or starving, let us put the money and man-hours being devoted to MOP into humanitarian efforts here on Earth. Or let us spend the money to clean up the pollution on our planet. Who cares about alien life?

You should care. Every thinking human being should care. Not because awareness of alien intelligence would help us to prepare for a possible invasion from space. That is the realm of science fiction. Rather, it would give us a sense of our place in the universe, our "status . . . in the cosmic pecking order," to quote renowned science and science-fiction writer Arthur C. Clarke. Is intelligent life unique to *Homo sapiens*, or is it rather commonplace in the cosmos? If it is commonplace, communication with alien life would provide us with insights into the diversity and richness of life, the alternate pathways that evolution might take, what it means to be human. We might learn of new technologies, medical advances. . . .

And if life is unique to Earth, if indeed we *are* alone, that is important to know as well. To quote Clarke again, "That is the most awesome possibility of all. We are only now beginning to appreciate our duty toward the planet Earth: If we are indeed the sole heirs to the galaxy, we must also be its future guardians."

In terms of expenses, the ten-year Microwave Observing Project is rather cheap. The Apollo moon shots, between 1962 and 1973, cost NASA $24 billion. The $13.5 million earmarked for MOP over the next year is *less than one tenth of one percent* of NASA's $15 billion annual budget. "When you factor in the consequences of success," astronomer Frank Drake says, "this could be the biggest bargain in history."

Over the past several centuries, we've come a long way in our thinking on this subject. On February 17, 1600, a man named Giordano Bruno was burned at the stake for saying that other worlds with intelligent life existed. Today, scientists around the globe are joining a crusade to find this intelligent life, to communicate with it, and to share knowledge with it.

Floating on a Sea of Rock

Earthquakes are the most destructive of all natural disasters. In 1201 more than a million people perished in an earthquake in northern Egypt. No hurricane, tornado, or volcanic eruption in the history of mankind has killed as many people. Over the past 2,000 years, three quarters of the world's major natural disasters have been earthquakes. Death and destruction occur not only from the quakes themselves but from fires caused by severed electrical wires and ruptured gas mains and from diseases caused by broken sewage lines that contaminate drinking water. Earthquakes that occur under water (and most do) may also give birth to huge tidal waves, called *tsunamis*, that batter populated coastal areas and cause severe flooding. In this essay we will explore earthquakes—what they are, where and why they happen, and what can be done about them.

Bad Vibrations

An earthquake is a shaking of Earth's crust accompanied by a release of energy. It occurs usually along a zone of weakness or fracture in the crust known as a *fault*. Not all earthquakes are violent or destructive. In fact, the vast majority are so weak that they go entirely unnoticed except by sensitive instruments designed to detect earth tremors. About one million earthquakes occur each year around the world—that's better than two every minute. Of these, however, only 3,000 are strong enough to move sections of crust, several hundred to move the crust significantly, and twenty or so to cause severe changes.

Why and Where Earthquakes Happen

From surface to center, Earth is composed of four distinct layers: the crust, a relatively thin solid and rocky layer; the mantle, also solid but with a kind of plasticity that allows it to flow slowly under pressure; the outer core, which is liquid; and the inner core, which is solid and metallic. The crust and upper mantle together, with a thickness of about 60 miles (97 km), are what is called the *lithosphere*. It is here that the world's most severe earthquakes originate.

What is interesting about the lithosphere is that it is not one continuous sheet of rock but is broken up into sections, or *plates*, like pieces of a jigsaw puzzle. To date, scientists have determined that this puzzle consists of about twenty pieces. Unlike a jigsaw puzzle, however, the lithospheric puzzle is constantly changing. The pieces are floating on a sea of sluggish, slow-flowing mantle. Under extremes of temperature and pressure from Earth's interior, these plates move. At their boundaries, which constitute major fault zones, they may move against one another, either vertically or horizontally. This sudden scraping of one plate against another is responsible for the vast majority of earthquakes that occur on Earth. They are called *interplate earthquakes*.

The earthquakes that periodically shake California are of this type. They are the result of two major plates—the Pacific Plate and the North American Plate—moving against each other. (The boundary between these plates in California is known as the San Andreas Fault.) If these plates slid smoothly and easily, there would be no problem. This is not the case, however; plates have a tendency to get stuck. Only after a buildup of considerable pressure or stress will they move, usually with a quick, snapping motion and the sudden release of enormous amounts of energy. This abrupt movement and energy release is an earthquake.

Not all earthquakes occur at the boundary between crustal plates. Three of the largest earthquakes in the history of the United States occurred along the Mississippi River near New Madrid, Missouri, an area well within the North American Plate. Though not at the boundary between plates, it is at the juncture of three active faults that were formed about 600 million years ago by various geologic forces. Scientists believe that these faults are being squeezed together by the movement of adjacent plates, causing stress—and eventual movement—within the fault zone. Earthquakes of this type, which do not occur at plate boundaries, are called *midplate earthquakes*. Major earthquakes in China are "midplates."

On the average, plates move one to several inches per year. The theory and study of plate movement is known as *plate tectonics*.

Locating Earthquakes

When you drop a pebble into water, a disturbance is created. This disturbance generates waves, which move outward from the source in all directions. Sudden fracturing and moving of Earth's lithosphere also creates disturbances, which generate waves that move outward in all directions. These waves are called earthquake waves, or *seismic waves*. There are three different kinds of seismic waves: *P waves*, *S waves*, and *L waves*.

P waves—also called *primary*, or *compression waves*—are caused by the squeezing together and spreading out of rock material as the lithosphere moves. They travel through all materials—solid rock, liquid rock, ocean water, and air. They travel the fastest of the three types of seismic waves and cause buildings to shake vertically.

S waves—also called *secondary* or *shear waves*—are caused by the shearing of rock material as the lithosphere moves. They travel only through solid materials and at about half the speed of P waves. They cause buildings to shake horizontally.

L waves—are also called *surface waves*. Both P and S waves are called body waves, because they travel through the body of Earth. When they reach the surface, they propagate L waves. Such waves travel only at the surface—they are not body waves. In their mode of travel, they are similar to water waves. They travel the slowest of the three wave types.

Earthquake waves are detected by an instrument known as a *seismograph*. This device can be used to locate the origin of an earthquake. (Actually, it locates the epicenter, which will be discussed later.) Remember that earthquake waves travel at different speeds: P waves produced by an earthquake will reach the seismograph first, S waves next, and L waves last. Seismologists record the difference in arrival time between these waves. By applying this difference to a special kind of graph, called a time-travel graph, they are able to determine how far away an earthquake is from the seismograph.

This is not enough information, however, to *locate* an

earthquake. As shown in Figure 17, an earthquake that occurred 700 miles from recording station A may have occurred at *any point* on the drawn circle whose center is at station A and which has a 700-mile radius. How, then, can the location of the quake be determined?

Figure 17

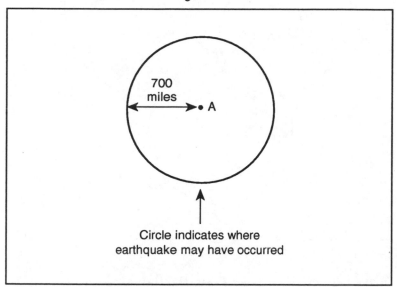

Circle indicates where
earthquake may have occurred

Triangulation, my dear Watson! If seismographic data are collected from at least *three separate stations*, and the distance to the quake from each one is calculated, the exact location of the quake can be determined. Figure 18 illustrates this concept.

The method of triangulation is used to determine an earthquake's *epicenter*, not its origin. The epicenter is the point on the surface of Earth *directly above* the origin of the quake. You see, earthquakes do not normally originate on the surface but below it. The point of origin of an earth-

Figure 18

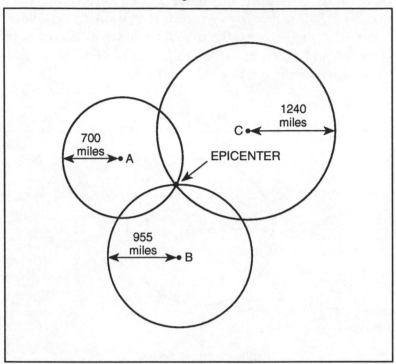

quake is referred to as its *focus*. Earthquakes vary greatly as to the depth of their foci. Shallow-focus earthquakes occur within about 40 miles (65 km) of Earth's surface. Intermediate-focus earthquakes occur at depths between 40 and 200 miles (65 and 320 km). Deep-focus earthquakes range from 200 to 400 miles (320 to 645 km) down. Below 400 miles, earthquakes do not occur. The high temperature and pressure at these depths cause rock to bend, like soft plastic, rather than to fracture.

The greatest damage is usually caused by shallow-focus earthquakes; they reach the surface more quickly, before much of their energy has been dissipated. The earthquakes of California are of this type.

Measuring Earthquakes

How "big" is an earthquake? That depends on what you're looking for: the energy it releases or the damage it does. An earthquake occurring in a barren desert will do less damage than one occurring in a major city, regardless of the punch that it packs. One occurring deep underground will do less damage than one occurring near the surface. Scientists distinguish between punch and damage in measuring an earthquake. An *intensity scale* expresses the severity of an earthquake in terms of damage, or the effects it produces on its surroundings. A *magnitude scale* expresses severity solely in terms of punch, or the amount of energy the quake releases.

The intensity scale used by geologists today is called the *Mercalli scale*, after Giuseppi Mercalli, an Italian geologist. It replaced an earlier scale in 1902 and was modified in 1931 to include tall buildings, cars and trucks, and underground pipes. It does not rely on instrumentation but on directly observable effects, making it a less-than-precise system of measurement. The scale consists of twelve levels of intensity, designated by Roman numerals:

Level I is ordinarily detected only by instruments.

Level II is felt by a few persons.

Level III is quite noticeable, but the tremors are not usually associated with an earthquake.

Level IV is felt by mostly everyone indoors and can rattle a few windows.

Level V can break windows and cause buildings to tremble. People may become frightened.

Level VI is felt by all and can move heavy furniture. Damage is slight.

Level VII may cause more serious damage to poorly built structures.

Levels VIII and IX are destructive to most buildings, with general panic among the population. Underground pipes may break.

Levels X and XI are generally catastrophic, with total panic among residents. The ground cracks in many places.

Level XII is total destruction to almost all structures. Surface waves are visible, with numerous ground cracks, landslides, and rockfalls. Objects are thrown into the air. It doesn't get any worse than this.

The magnitude scale in general use by geologists is the *Richter scale**, developed in 1935 by Charles F. Richter, of the California Institute of Technology. It is based on the height, or *amplitude*, of earthquake waves as measured on a seismograph. The scale, consisting of numbers ranging from 0 to 9, is logarithmic, which means that each increase of one whole number represents a *tenfold* increase in wave amplitude. Since energy content happens to increase *3.2 times* as rapidly as wave amplitude, the energy release of a quake increases *thirty-twofold* (10×3.2) for each whole-number increase on the scale. In other words, an earthquake of magnitude 6.0 has a wave amplitude *100 times* greater (10×10) than a magnitude 4.0 and an evergy release about *1,000 times* greater (32×32). Here is a sampling of Richter scale values and their probable intensities:

Magnitude	Intensity
2.5	Felt only by instruments and a few people. Called micro-quakes. (Levels I and II on the Mercalli scale.)
3.5	Felt by many people. Windows can rattle or break. (Levels IV and V on the Mercalli scale.)

---

*A newer scale, called the seismic moment scale, is coming into use. It employs different criteria for measuring earthquake magnitude and is believed by many geologists to be more accurate than the Richter.

4.5	Felt by all; many become frightened. Damage is minimal to well-built structures, considerable to poorly built ones. Energy release equivalent to a standard A-bomb blast (Levels VI and VII on the Mercalli scale.)
6.0	Ranges from alarm to panic among the local population. Ground cracks noticeably. Considerable damage to many structures. Energy release equivalent to a small H-bomb blast. (Levels VIII and IX on the Mercalli scale.)
7.0	A major earthquake—about ten occur each year. Ground is badly cracked. Most buildings are destroyed or badly damaged. Energy release could heat New York City for one year. (Level X on the Mercalli scale.)
8.0 and above	A great earthquake—occurs every five to ten years. Damage is severe or total to all buildings and bridges. Energy release could heat New York City for at least thirty years. (Levels XI and XII on the Mercalli scale.) Examples include the three earthquakes in New Madrid, Missouri, mentioned earlier. The largest magnitude earthquake ever recorded was an 8.9.

Since 1900, there have been about forty earthquakes that have registered—or would have registered, based on damage description—8.0 or greater on the Richter scale. In the United States a major earthquake happened on October 17, 1989, in the San Francisco area, suspending play of the baseball World Series between the San Francisco Giants and the Oakland Athletics. It registered 7.1 on the Richter, killed 67 people, and caused $6 billion in damages. More recently, on July 28, 1992, an earthquake in California's Mojave Desert registered 7.5 on the Richter. It was the most powerful California quake in 40 years. Luckily, there was only one fatality.

In terms of lives lost during earthquakes, the country to suffer most dearly over the course of history is China. A quake in 1556 in central China killed more than 800,000 people. As recently as 1976, a quake in northeastern China killed up to three quarters of a million people. (Figures on this earthquake vary greatly.) Can the tragic loss of life and property damage caused by major earthquakes be prevented or lessened?

Minimizing the Effects of Earthquakes

If geologists could predict a major earthquake before it occurs, measures could be taken to evacuate the affected area or at least the buildings and bridges where collapsing parts exact a heavy toll on human life. Though we are not yet at the point where earthquake prediction is at all reliable, especially in the short term, several methods have been used with some success:

Foreshocks. Major earthquakes are often foreshadowed by smaller quakes weeks or months before they occur. Unfortunately, these foreshocks may occur *years* before, making accurate timing of the quake difficult.

Uplifting. The crust may heave or uplift over a period of months to years before a major earthquake. As with foreshocks, it is difficult to make an accurate time prediction by studying uplifts.

Radon. The concentration of radon, a radioactive gas, has been found to increase in deep-well water prior to an earthquake. These findings are promising but not always conclusive.

Electrical conductivity. Rock and soil have demonstrated a significant reduction in the resistance to the flow of electricity prior to an earthquake. These results are also promising.

Animal behavior. The following excerpt is from an earthquake study delegation in China, 1979: "Some instances noted [before the quake] were of chickens refusing to enter their coops, cows breaking their halters and escaping . . . rats appearing to behave as though drunk." Seismologists have noticed dogs barking during the foreshocks of quakes. Nonetheless, the practical value of animal behavior in predicting earthquakes is questionable at best. (If Rover scratches, check for fleas rather than an earthquake.)

Seismic gaps. In some areas, earthquakes have a history

of occurring at fairly regular intervals. (In California, catastrophic quakes are predicted every fifty to a hundred years.) Seismologists keep track of these areas and note when earthquakes are due—or overdue. They have had moderate success over the mid- and long term in predicting earthquakes by this method.

Geyser Eruption. The Old Faithful geyser, in Calistoga, California, has shown a change in its eruption pattern before major earthquakes. The geyser normally shoots a 60-foot (18-meter) plume of water and steam into the air on a regular schedule. Carnegie Institute scientists Paul Silver and Nathalie Valette-Silver reported in the September 4, 1992, issue of *Science* that the geyser showed an irregular eruption pattern one to three days before three major earthquakes that shook northern California from 1975 to 1993. The scientists do not know what causes the geyser to change its eruption pattern, but the results are very promising, especially in the area of short-term warning. Other geysers in earthquake-prone areas are presently being monitored.

Smart Buildings. On the corner of a crowded Tokyo street, the world's first antiearthquake building stands—an eleven-story structure that has reduced seismic vibrations by as much as 80 percent. Unlike other buildings designed to withstand quakes, it does not attempt to resist them but to compensate for them. Using sliding weights to counterbalance the building as it tilts, flexible cables to pull it back to center when it sways, and jet thrusters to steady it when it vibrates, the building represents the state of the art in earthquake-prevention technology. The problem is that it may take another ten years to develop the technology to make the present technology cost-effective.

To this point, earthquakes have been given rather a bad rap—which indeed they deserve—as instruments of death and destruction. On the other side of the balance sheet, however, there are a *few* positives worth mentioning. For one, it is through the movement of seismic waves that we have learned much about Earth's interior. We have learned that it consists of four layers, one of which is liquid in nature. We have also been able to calculate the density of each layer. (Seismic waves travel faster through denser materials.)

Also, many historians believe that the Trojan War was ended by an earthquake, not by a large wooden horse with soldiers hidden in its belly. The city of Troy had been under siege for more than ten years by the Greeks, who were unable to penetrate its mighty walls. Scholars suggest that these walls finally may have been toppled by an earthquake rather than as a result of a horse wheeled through its gates as an offering of peace.

A bit of "faulty" history, if you will.

A Tropical Breeze
Out of Control

Camille, Hugo, Andrew, Diane, Elena, Betsy—each name carries with it chilling images of death and devastation. They are the names of hurricanes that have hit areas of the United States in the latter half of the twentieth century. Hurricanes are the only natural disasters with their own names. They are also the most powerful and destructive of all weather systems. Each year they do billions of dollars' worth of damage.

If we define a storm as any atmospheric disturbance accompanied by strong winds, then a hurricane is a *tropical storm*. As the name suggests, such a storm occurs, or at least forms, in a warm region of the world. Storms that form in cooler regions are called *extratropical storms*. Thunderstorms, which occur often in the continental United States, giving rise to snowstorms, blizzards, and tornados, are extratropical storms.

A hurricane is a low-pressure system, as are all storms;

in other words, the air at the center of the system has a lower pressure than the air outside it. A low-pressure air system is called a *cyclone*. Meteorologists will therefore refer to a hurricane as a *tropical cyclone*. In fact, in the South Pacific and in the Indian Ocean, hurricanes are called tropical cyclones. In the North Pacific, they are called *typhoons*.

Are all tropical storms or tropical cyclones hurricanes? Definitely not. The distinction is based on wind speed. If the maximum sustained winds within a tropical storm system are below 39 mph (63 km/h), the system is called a *tropical depression*; if they are between 39 mph and 74 mph (119 km/h), it is called a tropical storm; at 74 mph and greater, the system is a hurricane. Very simply, a hurricane is a high-speed tropical storm.

Where Hurricanes Are Born

As already noted, tropical storms are born in warm regions. These regions are at or near the equator, where rays of sunlight strike Earth most directly. Figure 19 shows where tropical storms form. On the average, about one hundred tropical storms form around the world each year; of these, about two-thirds develop into hurricanes.

As Figure 19 indicates, most hurricanes form in the western North Pacific Ocean. Of the hurricanes that strike the continental United States, the west-coast variety are born in the eastern Pacific Ocean. An average of sixteen tropical storms form each year in the eastern Pacific, off the coast of Central America and Mexico. They are on the shy side, rarely hitting land; and if they do, they usually are greatly weakened by the cold water off California. Little attention is paid to them, although occasionally one of destructive strength does strike land. Hurricanes Iwa (November 1982) and Iniki (September 1992), both of which

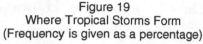

Figure 19
Where Tropical Storms Form
(Frequency is given as a percentage)

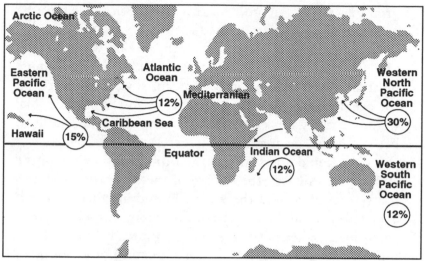

struck the islands of Hawaii, are two examples. Iniki was the state's costliest natural disaster, causing $1.2 billion in damages and destroying more than 8,000 homes. Iwa was the first hurricane to hit Hawaii since 1959.

The problem hurricanes in the United States, except for Hawaii, are the east-coast variety. They are born either in the Atlantic Ocean, the Gulf of Mexico, or the Caribbean Sea. More than half a dozen may strike the United States in a given year, usually from August to October. The most recent of these was Hurricane Andrew, which struck the coasts of Florida and Louisiana in August 1992. It caused 33 deaths, destroyed or damaged 136,000 homes, left 300,000 people homeless, and caused about *$30 billion* in damage. It was the costliest hurricane in United States history.

What can cause moving air to become so violent?

How a Hurricane Happens

Fewer than 10 percent of tropical atmospheric disturbances grow into storms. In order for that to happen, there must be a combination of several ingredients:

(1) A large expanse of warm ocean water—80° F (27° C) or higher—must be present. The warm water must be at least 200 feet (61 meters) deep, since storms stir up the ocean, bringing water from below to the surface. The warm surface water evaporates, causing the air above it to become moist, or humid.

(2) Humid air is lighter than dry air and rises in the atmosphere. Air near the surface of the water is sucked into the system to replace the warm, humid air that is rising. This causes wind. Due to Earth's rotation, this wind spirals toward the center of the system, forming a cylindrical column of rising air around the storm center. In the Southern Hemisphere, the spiraling is clockwise; in the Northern Hemisphere, including the United States, it is counterclockwise. In fact, *all* storm systems in the United States have a counterclockwise wind pattern. (The effect of Earth's rotation that causes this spiraling is called the *Coriolis effect.*)

(3) As the humid air rises, it cools and condenses. Condensation is a heat-releasing process. The heat that is released warms the surrounding air. Warm air is lighter than cool air. The warmed air, therefore, continues to rise, cool, and condense. This condensation releases further heat energy, continuing the cycle. As long as there is warm, humid air near the surface to "feed" the tropical storm system, it will continue to grow in size and intensity, ultimately becoming a hurricane. *The heat energy released by a hurricane in one day can supply all of our nation's electrical needs for about six months.*

Hurricanes weaken and die when they move over cooler

bodies of water or onto land, losing their supply of warm, humid air.

(4) The preexisting winds—those not caused by the hurricane—should be coming from nearly the same direction and at nearly the same speed at all altitudes so they will not tear the storm apart. It is the preexisting winds that largely determine the path that the hurricane takes.

Most hurricanes have a life span of from nine to twelve days.

Figure 20
Structure of a Hurricane

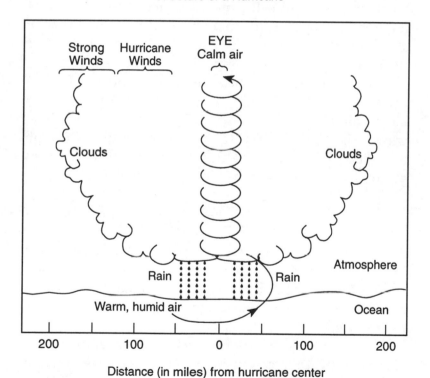

Distance (in miles) from hurricane center

What a Hurricane Looks Like

All hurricanes have a similar structure, with similar features, as illustrated in Figure 20, a fully developed Northern Hemisphere hurricane. Notice that warm air is spiraling counterclockwise in a rising column. Winds are strongest around this column. Farther out, wind intensity decreases but may still be quite strong for several hundred miles. Rains are usually heavy. Curiously, the very center of a hurricane is a region of calm, clear, dry air. It is called the *eye* of the storm. In the eye, there is almost no wind; the sky may even be blue. When you are in the eye of a hurricane, you may think that it is over. However, within a short time hurricane conditions will return, this time with winds blowing from the opposite direction due to the spiraling flow of air. You will then be on the "back side," or "back end," of the hurricane.

A typical fully developed hurricane is about 300 miles (483 km) across; the eye averages about 15 miles (24 km) across. Some may crawl along at less than 15 mph, while others may scurry at speeds better than 75 mph (121 km/h). The speed of air *within* the hurricane varies greatly as well but may be as high as 200 mph (322 km/h).

Storm Surges

Although wind speed gets most of the publicity during a hurricane, more deaths and destruction result from storm surges that can knock down buildings and utility poles, rip boats from their moorings, and wash tons of sand away from beaches.

A storm surge is a large mound of water that rises above normal sea level, causing most of its damage when it hits land. Two factors within a hurricane are responsible for storm surges: high winds and a low-pressure center. The winds, which spiral toward the center of the hurricane, create a swirl of water that piles up as it is carried along with the storm. The low-pressure center of the hurricane contributes to this effect by not pushing down very hard on the water. In the deep, open ocean, the mound of water ends up flowing downward under its own weight, which prevents the storm surge from becoming very high: open-sea storm surges may be as high as 3 feet (1 m).

Near land, however, where water is shallow, the ocean floor prevents the mound of water from flowing downward. Instead, it piles up higher and higher, generally to one side of the eye. Storm surges that strike land where the off-shore water is shallow for long stretches—such as the Gulf of Mexico—can be in excess of 25 feet (7.6 m). Hurricane forecasters consider the New Orleans area to be the most susceptible to such dangerous surges. In fact, the worst storm surge ever recorded in the United States came ashore with Hurricane Camille on August 17, 1969, at Pass Christian, Mississippi. Water rose twenty-four feet above normal, flooding the surrounding area. More than 5,500 homes were destroyed, with another 12,500 sustaining major damage. In all, 256 people died. If Camille's path had taken it just a bit farther west, its storm surge would have flooded downtown New Orleans, leaving the city under 20 feet (6 m) of water. Devastation would have been incalculable.

How Hurricanes Are Classified

In the early 1970s, Herbert Saffir, a consulting engineer, and Robert Simpson, then director of the National Hurricane Center, developed a scale for classifying hurricanes. It is known as the Saffir-Simpson hurricane damage-potential scale, or the Saffir-Simpson scale. It takes into account three elements of a hurricane: atmospheric, or barometric, pressure, sustained wind speed, and storm surge. The table below summarizes the Saffir-Simpson scale.

Category	Barometric Pressure (inches)	Wind Speed (mph)	Storm Surge (feet)	Damage Potential
1	>28.9	74-95	4-5	Minimal
2	28.5-28.9	96-110	6-8	Moderate
3	27.9-28.5	111-130	9-12	Extensive
4	27.2-27.9	131-155	13-18	Extreme
5	<27.2	>155	>18	Catastrophic

Hurricane Andrew, the costliest in U.S. history, was a Category 4. So was Iniki, which devastated the Hawaiian island of Kauai. Only two Category 5 hurricanes have ever hit the U.S. since recordkeeping began: the 1935 Labor Day hurricane that hit the Florida Keys, killing six hundred people, and Hurricane Camille, which devastated the coast of Mississippi in 1969.

The 1935 hurricane did not have a name. When *did* hurricane naming begin? And who names them?

Naming of Hurricanes*

In 1941, a popular novel called *Storm* by George R. Stewart, featured a meteorologist who labeled storms using women's names. The idea caught on, and starting in 1953 hurricanes were officially labeled with women's names. In 1979, men's names were added to the list as well as Spanish and French

---

*Hurricane naming is for Atlantic Basin hurricanes—those that form in the Atlantic Ocean, Caribbean Sea, or Gulf of Mexico.

names. A storm is named when its winds reach tropical storm strength—39 mph.

Hurricanes within a particular year are given names in alphabetical order; the names are chosen in advance over a repeating six-year cycle. The following chart lists the cycle of names from 1992 to 1997.

1992	1993	1994	1995	1996	1997
Andrew*	Arlene	Alberto	Allison	Arthur	Ana
Bonnie	Bret	Beryl	Barry	Bertha	Bill
Charley	Cindy	Chris	Chantal	Cesar	Claudette
Danielle	Dennis	Debby	Dean	Diana	Danny
Earl	Emily	Ernesto	Erin	Edouard	Erika
Frances	Floyd	Florence	Felix	Fran	Fabian
Georges	Gert	Gordon	Gabrielle	Gustav	Grace
Hermine	Harvey	Helene	Humberto	Hortense	Henri
Ivan	Irene	Isaac	Iris	Isidore	Isabel
Jeanne	Jose	Joyce	Jerry	Josephine	Juan
Karl	Katrina	Keith	Karen	Klaus	Kata
Lisa	Lenny	Leslie	Luis	Lili	Larry
Mitch	Maria	Michael	Marilyn	Marco	Mindy
Nicole	Nate	Nadine	Noel	Nana	Nicholas
Otto	Ophelia	Oscar	Opal	Omar	Odette
Paula	Philippe	Patty	Pablo	Paloma	Peter
Richard	Rita	Rafael	Roxanne	Rene	Rose
Shary	Stan	Sandy	Sebastien	Sally	Sam
Tomas	Tammy	Tony	Tanya	Teddy	Teresa
Virginie	Vince	Valerie	Van	Vicky	Victor
Walter	Wilma	William	Wendy	Wilfred	Wanda

*retired

After a hurricane has caused great damage, its name is retired—kind of like retiring a great athlete's jersey number. The following "superstars" have been retired (in chronological order): Camille (1969), David (1979), Frederick (1979), Allen (1980), Alicia (1983), Elena (1985), Gloria (1985), Gilbert (1988), Joan (1988), Hugo (1989), Bob (1991), and, of course, Andrew (1992).

Improved Forecasting

Since the turn of the century, the cost of hurricane damage has steadily increased, but death tolls have steadily decreased. Between 1900 and 1909, 8,100 people in the United States died from hurricanes. In the 1950s, the number was down to 750 and in the 1980s, only 161. The greatest single killer was one that hit Galveston, Texas, in 1900, killing 6,000. None of the ten worst killers occurred after 1957. Why not?

Since the turn of the century, hurricane forecasting technology has vastly improved. On April 1, 1960, the first satellite photos of a storm system were taken. Satellite reconnaissance, among other advances such as computer projection models, has led to improved tracking of storms. As a result, the National Weather Service's Hurricane Center in Coral Gables, Florida, is better able to alert people of impending storm systems, allowing them more time to either evacuate or "batten down the hatches."

Where Has All the Ozone Gone?

When an electric motor is running, it gives off a characteristic pungent odor. This is the smell of *ozone*, a pale blue, highly poisonous gas. Although present in only trace amounts in our atmosphere, ozone has gotten quite a bit of media coverage through the 1970s and 1980s. Exactly what is ozone and what makes it so newsworthy?

Chemically, ozone is a very simple molecule, almost identical to that of ordinary oxygen gas. Both are composed solely of the element oxygen, but whereas a molecule of oxygen gas is made of two oxygen atoms (0_2), an ozone molecule is triatomic (0_3).

Ozone can form in a variety of ways. Near Earth's surface, in the layer of air called the *troposphere*, it is generated by electric discharges such as lightning and the action of sunlight on nitrogen oxide air pollutants (released by burning fossil fuels). This tropospheric ozone is not desirable; it is referred to as *bad ozone*. It becomes a component of photochemical smog and is highly toxic. As small a concentration as 100 parts per billion of ozone molecules in the lower troposphere can irritate eyes, inflame mucous membranes, and make breathing difficult. Fortunately, there is very little bad ozone.

Beyond the troposphere lies the *stratosphere*. Ninety percent of the ozone enveloping our planet is found here, covering Earth in a blanket 10-20 miles (16-32 km) above the ground. This ozone is *good ozone*, for it protects the life below from an ultraviolet bath that would turn Earth into a barren and desolate rock. It is this layer of stratospheric ozone—the good ozone—that has been making headlines lately. Why?

Very simply, the good ozone, referred to as the *ozone layer*, is disappearing. To understand where it is going and why it is being depleted, we must look at what normally goes on in the stratosphere.

The ozone layer exists because ordinary oxygen (O_2) from the troposphere seeps up into the stratosphere, where it is bombarded by sunlight. Sunlight is a mixture of many different kinds of radiation—radiation of different frequencies and wavelengths (see "The Colors of Light"). Although most of the Sun's energy is emitted as yellow light, with a wavelength of 500-600 nm (nanometers, or 10^{-9} meters), a significant amount is radiated as ultraviolet (UV). UV radiation is of shorter wavelength than the visible spectrum and is therefore more energetic. It is UV radiation that causes sunburn. UV radiation also maintains the stratospheric ozone layer.

Ultraviolet radiation can be divided into several different bands, depending on wavelength. The most energetic UV has a wavelength below 240 nm. This radiation is readily absorbed by molecules of oxygen in the stratosphere, causing them to split into individual oxygen atoms. Each oxygen atom is very reactive and will soon find another oxygen molecule to bond with. The result is production of an ozone molecule:

$$(1)\ O_2 \xrightarrow{\text{shortwave}\atop\text{UV}} O + O$$

$$(2)\ O + O_2 \longrightarrow O_3$$

In causing the dissociation of oxygen molecules, the very dangerous shortwave UV is completely filtered out of sunlight. Were it permitted to shower down upon us, this radiation would cause severe damage to RNA, DNA, and protein in cells. These organic molecules are the substances of life. In other words, the more energetic UV radiation would kill living things quite efficiently. A typical day at the beach would result in lethal sun poisoning and radiation burns.

The UV assault, however, does not end with shortwave absorption. UV radiation comes in increasingly longer wavelengths, called UV-C, UV-B, and UV-A. Although less energetic than the shortwave variety, UV-C and UV-B are, nonetheless, dangerous forms of ultraviolet. Fortunately, UV-C (240–290 nm) is completely filtered out of sunlight. So is most of the UV-B (290–320 nm). This time it is ozone, rather than ordinary oxygen, that absorbs UV radiation and renders it harmless. An ozone molecule is less stable than an oxygen molecule and can more easily be torn apart. When struck by this longer wavelength UV, it absorbs the energy and splits into a single oxygen atom and an oxygen molecule. The oxygen atom readily combines with another oxygen molecule (which are in abundance), reforming ozone. It is a cyclic process of absorption, splitting, and recombining in which, although no new substances are formed, most of the harmful UV is filtered out. Following is a depiction of these chemical events:

Figure 21

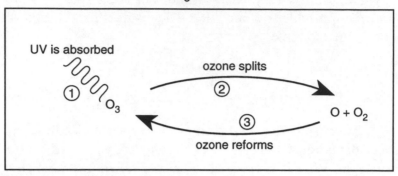

There are additional reactions that do convert ozone back to oxygen. These balance the reactions in which ozone is continually being formed from ordinary oxygen by short-wave UV. Here is one such reaction:

$$O + O_3 \longrightarrow O_2 + O_2$$

The overall picture is one in which a number of UV-induced chemical reactions occurring in the stratosphere accomplish two very important ends. First, a stable layer of ozone is maintained. Second, and more important, *life-threatening UV is removed or filtered from sunlight.*

Disappearing Act

Now we return to the question we asked a while back: Why is the ozone layer disappearing and where is it going? Simply put, human activities are discharging chemicals into the atmosphere that react with and destroy ozone. Harold Johnston, of the University of California at Berkeley, first warned of possible ozone depletion in 1971. His concern was over the new supersonic transport (SST), a high-flying, high-speed aircraft that was touted as the airplane of the future. Such aircraft as the *Concorde* would fly in the stratosphere and release nitrogen oxides during the burning of their fuel. Johnston presented evidence that these nitrogen oxides, producers of bad ozone in the troposphere, could efficiently destroy good ozone in the stratosphere. Fleets of hundreds of SSTs zipping through the stratosphere, spewing out their nitrogen oxide exhausts would, in Johnston's opinion, create a serious ozone deficiency. Fortunately, the commercial success of SSTs was never realized; and besides, they flew too low in the stratosphere to cause serious ozone damage.

But Johnston's work did create an awareness of the impact that human activities could have on the ozone layer. So when the space shuttle program came along, with its promise of perhaps one shuttle flight a week, environmental-impact studies were conducted. The studies showed that one gas released by the space shuttle, hydrogen chloride (HCl), could pose a threat to the ozone layer. This was the first time that the element *chlorine* was implicated in possible ozone depletion. It certainly would not be the last.

Like the SST scare, the shuttle scare proved to be groundless. Weekly shuttle flights never materialized and there simply was not enough HCl being released into the stratosphere to have any significant effect on the ozone. Tragically, the same cannot be said for another group of chemicals, the *chlorofluorocarbons*, or CFCs.

CFCs derive their name from the fact that they are composed of the elements chlorine, fluorine, and carbon. There are a number of different CFCs, but the two most important are called F-11 and F-12. Here are the molecular formulas of each:

$$Cl - \overset{\displaystyle Cl}{\underset{\displaystyle F}{C}} - Cl \qquad Cl - \overset{\displaystyle Cl}{\underset{\displaystyle F}{C}} - F$$

F-11   F-12

C = carbon atom
Cl = chlorine atom
F = fluorine atom

CFCs were developed in the 1920s as refrigerants—first for refrigerators and later for air conditioners. (DuPont, the major producer of CFC refrigerants, marketed theirs under the brand name Freon.) Because CFCs were nonflammable, nontoxic, cheap to manufacture, and chemically inert, they also seemed perfect for use as propellants in aerosol, or spray cans. In 1950 the first aerosol cans using CFC propellants were marketed. By 1973 the number of these cans being produced yearly had grown to six billion (half in North America).

Indeed, CFCs were almost too good to be true—ideal as both refrigerants and propellants. But such widespread dissemination of a wholly new and man-made gas into the atmosphere certainly warranted some investigation. So Jim Lovecock, a British scientist, set about building a very sensitive CFC detector, called a "sniffer." In 1971 and 1972 he took the instrument on a sea voyage from Britain to Antarctica and back again. The sniffer detected traces of CFCs to the tune of a few to a few dozen parts per trillion (ppt). His conclusion was that they posed "no conceivable hazard" to the environment.

Lovecock was wrong, as two other scientists, Sherry Rowland and Mario Molina, would point out a year later. Rowland and Molina knew, as did Lovecock, that nothing happened to CFCs in the troposphere. But they went a step further. Sooner or later, they reasoned, these CFCs would work their way up to the stratosphere. (Subsequent studies have shown that this rise from ground level to the stratosphere takes from six to thirteen years.) And it was here that UV-C radiation would tear apart the CFC molecules, releasing chlorine atoms:

$$CCl_3F \longrightarrow Cl + CCl_2F$$

(This shows the dissociation of an F-11 molecule: a similar dissociation occurs with F-12 molecules.)

It is release of these free chlorine atoms that presents a

major problem for the ozone layer. Free chlorine atoms are like a cancer in the way they attack and destroy ozone. They are a thousand times more likely to react with an ozone molecule than with any other compound in the stratosphere. It is predicted that one chlorine atom can destroy up to 100,000 ozone molecules once it is dissociated at the stratospheric level and before it gets locked up in some less reactive form. Here is how this happens:

$$Cl + O_3 \longrightarrow ClO + O_2$$
$$ClO + O \longrightarrow Cl + O_2$$
$$\uparrow$$
(from UV assault on O_2 and O_3)

Net reaction:

$$O_3 + O \longrightarrow O_2 + O_2$$

As you can see, the chlorine atom effects the change from O_3 to O_2 yet is itself unchanged in the end. A substance that brings about chemical change without being changed itself is called a *catalyst*. Because chlorine is a catalyst, very little can go a long way in destroying ozone.

Rowland and Molina's findings opened up a real can of worms (or would that be "can of CFCs"?). Scientists around the world began measuring and monitoring CFC and ozone levels in the atmosphere. Fluctuations in UV radiation were superimposed upon these findings. The chemistry of the stratosphere became a science unto itself, involving more than fifty compounds and maybe one hundred fifty chains of chemical reactions. Rowland and Molina themselves predicted that 20 to 40 percent of the ozone shield would be destroyed within thirty years. Hardly a week went by without a major magazine or newspaper story about ozone depletion. And at every turn the CFC producers disputed the evidence. They called the environmentalists "doomsday watchers," claiming their figures to be either gross exaggerations or pure nonsense.

The battle raged through the 1970s. Finally, in 1976 a long-awaited National Academy of Sciences (NAS) study concluded that even if CFCs were held to 1973 levels, there would be a 6 to 7.5 percent long-term reduction of stratospheric ozone. Since the rule of thumb is that every 1 percent reduction in ozone produces a 2 percent UV-B increase, this equates to a 12 to 15 percent increase in UV-B. Largely as a result of these findings, manufacture and use of CFCs for aerosol propellants was banned in the United States in 1978.

The spray-can battle had been won by the environmentalists. The war against CFCs, however, was not yet over. DuPont and other chemical companies continued to aggressively market their product to the electronics industry, where it was—and still is—used as a solvent in cleaning silicone microchips. CFCs are also widely used in the process of blowing plastics into foam for home insulation, styrofoam cups, and packaging materials. Add to this the leakage from refrigerators and air conditioners, and it is obvious why CFC levels continue to climb. Lovecock's measurement of maybe a dozen parts per trillion of CFC molecules in the atmosphere has skyrocketed to over 600 ppt.

And CFCs are so terribly durable. It is estimated that F-12 molecules can survive for up to *one hundred fifty years* in the upper atmosphere before being broken down. CFCs released in the 1960s will still be floating around and harming our planet well into the twenty-first and possibly the twenty-second century—even if CFC emissions are halted today.

A Hole in the Sky

If there was ever any doubt that the "doomsday" scientists and their calculations might be wrong or at least overstated, a 1982 finding put these doubts to rest. In that year, a startled team of scientists from the British Antarctic Survey found a literal "hole" in the ozone layer at Halley Bay, Antarctica. The hole seemed most pronounced in the springtime, where *more than 20 percent* of the ozone over Antarctica disappeared.

Subsequent studies showed the hole to be getting worse with each passing year. Finally, in 1987, NASA put together a massive international expedition to explore the ozone hole. Called the Airborne Antarctic Ozone Experiment (AAOE), it sent two aircraft on more than two dozen flights deep into the hole to measure at different altitudes not only the ozone levels but also levels of CFCs and other ozone scavengers.

Findings of the AAOE confirmed the atmospheric scientists' worst fears: a huge springtime hole the size of the United States and as deep as the height of Mount Everest was centered more or less over the South Pole. Every conceivable chemical suspected of being involved in ozone depletion was present there in abnormal concentrations. As John Gribbin stated in his book *The Hole in the Sky*, "That expedition provided the final, incontrovertible evidence—the 'smoking gun'—showing that chlorine from CFCs was to blame." Overall, ozone depletion in the hole was greater than 50 percent, although at some altitudes there was almost no ozone at all.

The severe cold, unique to the stratosphere of Antarctica, causes polar stratospheric clouds to form. Inside these clouds, chemical reactions that occur nowhere else on Earth release chlorine atoms from a reservoir of chlorine compounds. They have little effect on the ozone until springtime, when the returning Sun triggers their ozone-eating capabilities.

It appears that ozone depletion is not confined to the

hole proper but is extending northward from Antarctica. Recently, much of the Southern Hemisphere has shown a springtime drop in ozone—as much as 20 percent in the stratosphere above New Zealand, southern Australia, and parts of South America.

And that's not all: since the 1987 NASA expedition, a smaller hole has been discovered over the North Pole as well. By 1993 the Total Ozone Mapping Spectrometer (TOMS) on the Nimbus-7 satellite showed record low levels of stratospheric ozone over much of the planet. Although some of the blame for this lies with the 1991 eruption of Mt. Pinatubo, which released a large amount of pollutants, CFCs are still the main culprit.

The prognosis is not good. In a 1987 global treaty hammered out in Montreal, twenty-seven technologically advanced nations agreed to reduce the release of CFCs by 50 percent—but not until the end of the twentieth century. The 1990 Clean Air Act of Congress called for the United States to phase out harmful chlorine compounds by the year 2000. Why the year 2000? Because it would allow DuPont to recover its investment in the technology. Unfortunately, the Rio de Janeiro Earth Summit, held in June of 1992 and attended by 178 nations, came up with no concrete deadlines for the banning of CFCs worldwide.

Meanwhile, solar UV, especially UV-B, continues its assault on life. It suppresses the human immune system, which is our body's defense against disease. It induces skin cancer, which is on the rise—especially the often-fatal *malignant melanoma*. It causes cataracts and retinal damage. So sensitive are living things to UV-B that it has been called *biologically active* ultraviolet. In places such as Punta Arenas, Chile, the world's southernmost city, people can safely go out for only brief periods of time during the day. The Reagan administration proposed that these people buy sunglasses and put on sunscreen to protect themselves. Now why couldn't the world's top scientists think of such a practical and simple solution?

Cattle, fish, agricultural crops, and phytoplankton (tiny green organisms that float near the ocean surface) are also adversely affected by UV-B. And they represent much of the world's limited food supply. Unfortunately, cattle and fish cannot put on sunglasses and sunscreen.

No one can accurately predict what catastrophes we are creating today for our children and grandchildren. Only time will tell. Environmentally safe substitutes for CFCs are being developed more rapidly than expected. Ironically, research is led by DuPont. Meanwhile, let us all enjoy what DuPont has provided for us so far—better living through chemistry. And let us remember what was said prophetically at the Stockholm conference on environmental awareness, Only One Earth, in 1972: "We have forgotten how to be good guests, how to walk lightly on the earth as its other creatures do."

What Happens to Us
When We Die?

The Body of B. Franklin,
Printer
Like the Cover of an Old Book
Its Contents Torn Out And
Stripped of Its Lettering and Guilding
Lies Here
Food for Worms.
But the Work Shall Not Be Lost,
For It Will, as He Believed,
Appear Once More
In a New and More Elegant Edition
Revised and Corrected
By the Author.

Ben Franklin wrote this epitaph for himself when he was twenty-two years of age. It clearly indicates his belief in reincarnation. Surprisingly, this same belief is held by 23 percent of all Americans, according to a 1982 Gallup poll. (A whopping 67 percent believe in some sort of life after death.) In Britain, the figure is even higher—28 percent as of 1979. This represents a 10 percent increase over ten years, and it prompted the manufacture of buttons that read "Reincarnation Is Making a Comeback."

The Near-Death Experience

Even the scientific community is getting into the act. Medical doctors as well as psychologists are investigating phenomena that point to the existence of a soul, a consciousness within our physical body that survives death. One of these phenomena, the near-death experience, was first described in 1975 by Dr. Raymond A. Moody, Jr., in his book *Life After Life*. Moody found striking similarities in the accounts of 150 people who had survived *very* close brushes with death—that is, people who had been clinically dead for up to several minutes before being revived. Based on these similarities, he constructed a composite near-death experience (NDE):

1. At the point of greatest pain or discomfort, they hear the doctor pronounce them dead.
2. Suddenly they find themselves out of their bodies, often floating overhead, watching people trying to save them. There is no longer any pain.
3. At about this time they start moving very rapidly through a long, dark tunnel.
4. They see spirits of relatives and friends who have already died.
5. A "being of light" appears and through nonverbal communication asks them to evaluate their

life. They see all major events of their life played back before them.

6. They come to a barrier separating earthly life from the afterlife.
7. They find they must go back into their bodies but do not want to return because the "light" is overwhelming them with feelings of joy, love, and peace.
8. Upon returning, they find words inadequate to describe how beautiful they felt while outside of their bodies. They no longer fear death and seem to better understand the purpose of living.

What is the significance of a near-death experience? Is it indicative of a soul separating from its physical body after death? Does an "entity" actually hover over the corpse?

According to a number of accounts, people who have survived NDEs are able to recount exactly what transpired while they were dead—what the doctors said to one another and the procedures that were performed on them are all accurately described. One disembodied soul even followed a nurse into another room and reported what that person did.

Incontrovertible proof of man's immortal soul, and the possibility of its being able to enter another body? Not so fast. Even Moody stopped short of such an explanation when he wrote, "Not one of these cases I have looked into is in any way indicative to me that reincarnation occurs. However, it is important to bear in mind that not one of them rules out reincarnation either."

If not flight of the soul to a cosmic plane, then what? A number of possibilities, with varying degrees of credibility, have been suggested. One popular explanation is that the near-death experience is nothing more than a hallucination induced by drugs administered to dying patients. This sounds credible, but it does not hold up under investigation. Melvin Morse, a renowned researcher in the field of NDEs among children, investigated thirty-seven youngsters who had been treated with "every kind of mind-altering medication known to pharmacology," including "anesthetic agents, narcotics, valium, thorazine, haldol, dilantin, antidepressants, mood elevators, and pain killers." Yet none of the children had anything remotely resembling an NDE. Furthermore, in researching the literature Morse found NDEs to be unique. No other hallucinations or visions produced by marijuana, psychedelics, or the like were similar to NDEs.

Endorphins are morphinelike substances produced by the brain to alleviate pain. They are responsible for the "runner's high" experienced by joggers. Some experts also feel they are responsible for NDEs. Although the stress of dying might increase endorphin levels in the brain, there is no evidence yet that the increases are significant or that they in any way trigger or promote near-death experiences.

In many NDE cases—drownings, cardiac arrests—the person's heart stops and breathing ceases. Perhaps the increase in carbon dioxide or lack of oxygen in the blood (called hypoxia) causes the NDE. Support for such a hypothesis can be found in the work of Dr. L. J. Medune, a professor of psychiatry at the University of Illinois School of Medicine. In tests conducted during the 1940s and 1950s he administered high concentrations of carbon dioxide to mentally ill patients for several minutes. This "Medune mixture" was aimed at correcting biochemical dysfunctions within the brain. The descriptions of many of his patients who breathed the gas were consistent with accounts of NDEs.

Ancient Egypt provides further confirmation for the hypoxia theory of NDEs. Three thousand years before the birth of Christ, priests and pharaohs willingly permitted themselves to be buried alive in casks sealed with wax. Eight minutes later, when they must have been precariously close to death by suffocation, the caskets were opened and they were revived. The point of this ritual? To bring on an NDE, which, in the eyes of ancient Egyptians, made the pharaoh a God-King.

But not all evidence supports the oxygen deprivation theory of near-death experiences. In one study, the blood gases of children who *had* NDEs were compared to hospitalized children who *didn't*. The results were conclusive. None of the patients experiencing NDEs were any more deprived of oxygen than the non-NDE children.

What does it all mean? In all likelihood, the close scrape with death caused by oxygen deprivation brought on the NDE and not oxygen deprivation per se. In studying NDEs, it seems that the one common factor, the ingredient necessary to the experience, is nearness of death. One must not only knock on death's door; one must open it and peek inside.

Carl Sagan, Cornell University astronomer and world-renowned science writer, has come up with a novel explanation for NDEs—memories of the birth experience: "Every human being, without exception, has already shared an experience like that of those travelers who return from the land of death: The sensation of flight, the emergence from darkness into light; an experience in which, at least sometimes, a heroic figure can be dimly perceived, bathed in radiance and glory. There is only one common experience that matches this description. It is called birth." Unfortunately, newborns do not have the mental capacity to remember such things; and if they did, they might not find the birth trauma to be all that pleasant.

Wilfield Pender, the father of neuroscience, performed

experiments in the 1940s that involved stimulating different areas of the human brain with an electric probe. When he probed the tissues surrounding the sylvian fissure, a groove of the brain between the right frontal and right temporal lobes, a remarkable thing happened. The patient had the sensation of leaving his body, of speeding down a tunnel, of seeing God, of talking with dead friends and relatives. Virtually every element of an NDE was experienced. More recently, a group of Chilean neurologists pinpointed the same area in the brain as the anatomical site of NDEs. According to these researchers, "near-death experiences were generated by neuron activity within the sylvian fissure."

Does this mean NDEs are merely physiological phenomena—the result of electrochemical impulses propagated along certain nerve pathways? Or have neurologists found within the sylvian fissure the "seat of the soul"? No one really knows, although feelings on both sides run very strong.

One thing that the pro-soulers have going for them is "the light." Although many features of an NDE can be explained by phenomena that are not paranormal, this is the one aspect of the experience that defies logical, scientific explanation. Just as the brain has all but shut down, one sees a brilliant light. And even more inexplicable are the feelings of love, caring, and abiding joy. These thoughts and emotions involve high-order brain functioning and are not the death flickers of an optic nerve or a final cerebral spasm.

It is interesting to note that the temporal lobe—possible "seat of the soul"—has already been linked to experiences of *déjà vu*, the feeling of having done something or been someplace before. It is experienced by many people at some time or another and then quickly dismissed as an illusion. But in some people the sense is very strong, and it involves not just a previous experience but a previous existence. In many instances, a person clearly remembers his or her name in that former life as well as specific places and events. The phenomenon is called *past-life recall*, and it may be the strongest evidence to date of reincarnation and the existence of an immortal soul.

Past-Life Recall

Ian Stevenson, M.D., is probably the world's best-known investigator of past-life recall. He has more than 2,500 cases of spontaneous recall on file in the Department of Parapsychology at the University of Virginia. In his classic work *Twenty Cases Suggestive of Reincarnation*, he extensively researched twenty of these cases from five different countries, all of them children ranging in age from one to seven years. In each case, great care was taken to insure accuracy and rule out possible fraud. For each case history, twenty to thirty facts that the child spontaneously remembered were investigated. Several sources had to corroborate the authenticity of each recollection. Every conceivable possibility short of reincarnation was considered. In all instances, Dr. Stevenson felt that rebirth was the only plausible explanation.

In rare instances, past-life recallers exhibit the phenomenon of *xenoglossy*, an ability to speak a language they have not been exposed to in their present lives. One man, treated by Joel Whitton, M.D., Ph.D., and author of *Life Between Life*, was able to communicate in two ancient languages that no longer exist. He also exhibited *xenography*, the ability to write in these strange languages.

Perhaps the most celebrated case of past-life recall occurred in the early 1950s to Virginia Tighe, a Denver, Colorado, housewife. Her recollections, however, were not spontaneous—they were induced by a phenomenon known as *hypnotic regression*. Whenever Ms. Tighe's neighbor, an amateur hypnotist, hypnotized her by candlelight, an early nineteenth-century Irish lass named Bridey Murphy emerged. Tighe's vivid recounting of her life as Murphy, in Cork, Ireland, was hailed for its accuracy of detail. These hypnotic regressions caught the imagination of the public and spawned worldwide debate on the subject of reincarnation. "Come as You *Were*" parties became fashionable.

One of the richest sources of reincarnation case histories comes to us from the files of Edgar Cayce, the greatest American psychic. From 1923 to 1945 people with various and sundry ailments and afflictions came to Cayce for "life readings." Over that twenty-two-year period he gave more than 2,500 readings, during which he would fall into hypnotic trances and regress to one or more of his clients' previous lives.

Cayce believed that past lives, or incarnations, exert a very real influence on present lives, that there is a carryover from one life to another as the soul evolves. Maladies being suffered now might well have their roots in unacceptable behavior of the past. Only through appropriate rebirths can the soul learn from its misdeeds and evolve to a higher level of consciousness. Gautama Buddha is said to have lived 550 lives over a 25,000-year period before achieving spiritual perfection.

This idea of compensation for past wrongdoings is called *karma*, and it guides the soul through its many incarnations. Karma is the driving force of reincarnation. Here are a few examples of how Cayce applied this notion in his life readings:

A college professor, born totally blind, had a life reading by Cayce, who outlined four previous incarnations—in America during the Civil War, in France during the Crusades, in Persia about 1000 B.C., and in Atlantis just before it submerged. It was in Persia that he set in motion the spiritual law that would result in his present blindness. He had been a member of a barbaric tribe whose custom was to blind its enemies with red hot irons, and it had been his role to do the blinding.

A boy who was a chronic bed wetter could not be cured by his parents, physicians, or psychiatrists. Finally, at age eleven, he was taken to Edgar Cayce. According to the boy's life reading, in a previous life he had been a minister at the time of the Salem witchcraft trials and was involved in punishing accused witches by tying them to a stool and dunking them into a pond—thus, the symbolic connection between water and bed-wetting. An eventual cure was effected by whispering to the boy as he slept that he was good and kind and would help many people.

Intriguing, to say the least, but is there any validity to past-life recall? When a subject under hypnosis recollects earlier existences, are they truly "memories of the soul"? Psychology professor Dr. Nicholas Spanos, of Carleton University, thinks not. According to him, past-life memories are "expectation-induced fantasies . . . directly influenced by the hypnotist's bias and the subject's own interests and concerns. . . . A person with an interest in Florentine art is likely to construct, under hypnosis, a minutely detailed life in Renaissance Italy."

Spanos has gone to great lengths to debunk seemingly airtight instances of recalled incarnations. In the Bridey Murphy case, for example, careful investigation of Virginia Tighe's past revealed that she once lived with an aunt of Scottish-Irish descent who captivated her with enchanting tales of the old country. Furthermore, one of Ms. Tighe's neighbors when she had lived in Chicago was a woman named Bridey Murphy Corkell.

Was this a case of deliberate deceit on the part of Virginia Tighe? Probably not. It was most likely a classic case of *cryptoamnesia*, a phenomenon in which bits and pieces of information, long forgotten yet stored away in the brain's subconscious, become the basis for full-blown fantasies. Under hypnosis, they emerge as past-life memories.

Whether real or imagined, past-life recall has proven to be a remarkable tool in the hands of trained psychotherapists. By hypnotically regressing clients to previous lives, past-life therapists are effecting cures for phobias, depression, asthma, migraines, and a host of other problems that resist traditional treatment. In many instances, merely reliving the experience cures the affliction. As one expert explains it: "A man with persistent neck pain sees himself guillotined in eighteenth-century France. And voilà! His neck pain disappears."

Today, membership in the California-based Association for Past-Life Research and Therapies is seven hundred strong and growing. Its members all practice past-life hypnotherapy, many with extraordinary success. Dr. Brian Weiss, a magna cum laude graduate of Columbia University and Yale Medical School, relates the story of a client who saw herself in a Middle Eastern country long ago, "riding in a wagon filled with wet straw. It overturned and she died, trapped and suffocating beneath the straw. After 'reliving' this episode, . . . her chronic asthma dissipated. For the first time in years, she can sleep through the night without waking up gasping for air."

Very possibly, there was no suffocation of a young girl under wet straw. It might only have been a "symbolic" story created by the unconscious to deal with the real problems and needs of the psyche. To most past-life therapists it doesn't matter. The important thing is that past-life regression has genuine therapeutic value. One therapist calls the technique "the single most effective psychotherapy tool I know."

Near death experiences. Past-life recall. There is even a man who has a scarlike lesion in the area of his left brain where he was allegedly shot and killed in a past life. What does it all mean? Does any of it prove the existence of an entity, a soul that survives death? Suggest, perhaps; prove, no—although the circumstantial evidence favoring reincarnation seems to be mounting. "After all," as Voltaire stated many years ago, "it is no more surprising to be born twice than it is to be born once."

Why Do We Grow Old?

Can the aging process be stopped? Can it be slowed? Why is it that the sea anemone, a brightly colored animal that lives rooted to the ocean floor and looks more like a flower than an animal, appears not to age? How does the California bristlecone pine manage to live for over 4,000 years?

In the early 1500s the Spanish explorer Ponce de León searched unsuccessfully for the mythical Fountain of Youth. In the famous Oscar Wilde novel, Dorian Gray's portrait aged while Gray himself retained his youth.

There are, of course, more scientific methods of fending off Father Time. Dr. John Kenneth Beddow, in his book *Stay Young—Reduce Your Rate of Aging*, recommends fasting every other day. He bases this regime on a pioneering experiment in the early 1970s in which the caloric intake of rodents was severely reduced. The diet restriction kept these animals youthful far longer than a control group and significantly prolonged their lives. How it retarded the aging process is still unclear.

In the United States during the early 1900s, John Romulus Brinkley developed a unique rejuvenation procedure: he inserted whole goat testes into human scrota. The operation was performed on roughly 16,000 patients. In 1948, the Swiss surgeon Paul Niehans began using cells from lamb fetuses for the purpose of rejuvenation. His daughter still runs the Niehans Cellular Therapy Clinic in the Swiss Alps.

A less invasive approach to lasting youth is practiced by certain lamas, or high priests, who live in remote Tibetan monasteries. They believe that the human body has seven energy centers. To keep the energy flowing properly throughout the body, one must outstretch the arms and spin. The whirling dervishes of India are famous for their spinning, which allegedly keeps them vibrant and youthful.

So what will it be, fasting, spinning, or scrotal surgery? Don't like the choices? Well, then, let us see what legitimate scientific research has uncovered about aging and its prevention.

The symptoms of aging are many: skin wrinkles, hair turns gray, brown "liver" spots appear on the skin, memory begins to fail, and sex drive diminishes. The ability to combat infection declines, and autoimmune responses such as arthritis increase, indicating immune-system failure. Between adulthood and retirement, as many as 50,000 brain cells die each day. By age sixty people have lost half their taste buds and about 10 percent of their brain cells. Endocrine secretions begin to dry up. The heart, kidneys, liver, and lungs fall far short of their once-optimum levels of performance. In general, the basic metabolism rate drops 50 percent from adolescence to age seventy. Actuarially, the probability of dying roughly doubles every eight years after puberty.

Why? What event or events occur in the body to bring about this catastrophic yet inevitable decline? There is no simple answer. Aging is an incredibly complex phenomenon. Gerontologists compare it to the elephant that six blind men are trying to understand—each one grasps a different part of the animal and comes away with a different impression of what it must look like. So it is with aging. Cell biologists view the process as one of random damage to our cells and tissues caused by a variety of toxic substances. Endocrinologists see aging as the result of hormonal changes and fluctuations. Geneticists believe that it is the turning on and off of specific genes at specific times in our lives—genetic clocks, if you will—that bring about aging. All are correct. Yet until we discover how the different theories fit together, we will not understand the true nature of the beast.

Random Damage

It seems as if the human body is programmed to live about 120 years. This is the maximum age granted us by our genes. Tragically, very few people attain this ripe old age. The oldest person now alive, a Frenchwoman, is 117 years old. The average life span of 75 years falls far below nature's "upper limit."

People die too young because they smoke, because they lie in the sun, because they eat and drink and breathe. In short, people die because the mere act of living produces toxic substances within the body that destroy it. Chief among these toxins is a class of highly reactive, electrically charged molecules or molecular fragments called "free radicals" (such as the negatively charged *hydroxyl* radical, OH^-). They are the vicious byproducts of reactions involving the oxygen we breathe, and there is no escaping them. Often, free radicals are necessary intermediates in normal metabolic pathways. The oxidation of fats, especially unsaturated fats, produces an abundance of these molecular assassins. High-energy radiation such as X rays as well as the less energetic ultraviolet rays found in sunlight produce free radicals in our cells. We even inhale them as they float in the air.

Each day these free radicals launch about 10,000 chemical attacks on every human cell. They wreak havoc on all known organic compounds—proteins, lipids, carbohydrates, and nucleic acids. According to a recent *Scientific American* article, Earl R. Stadtman, of the National Heart, Lung, and Blood Institute, estimates that as much as half of the proteins, including many enzymes, in elderly individuals might be oxidatively damaged—and thus nonfunctional. This would almost certainly promote senescence."

One type of chemical change wrought by free radical assault is cross-linking, the formation of chemical bonds between large molecules. It causes proteins to become stiff and less flexible, to lose their elasticity. *Collagen*, found in connective tissue, is the most abundant protein in the body. When it becomes cross-linked, wrinkling and a general aging of the skin results. Cross-linking of elastin proteins in artery walls decreases blood flow and is a contributing factor in coronary heart disease. Free-radical-induced cross-linking in lung tissue is a major factor in emphysema. There is little doubt that cross-linking of proteins is a major cause of the tissue deterioration we refer to as aging.

DNA, the stuff of which genes are made, also falls prey to attack by free radicals. DNA is the blueprint for living, the program that tells the body what to do and when to do it. Every time a cell in the body divides, the DNA must duplicate itself, so that each cell gets a complete set of blueprints. But duplication is not flawless. Each cell division produces errors, or mutations, in the newly replicated DNA. Radiation, free radicals, and a host of other chemicals greatly increase the probability of error. These errors accumulate in the DNA and translate into faulty messages that are sent to the machinery of the cell. Damage to DNA is now accepted as a principal factor in aging, especially damage to mitochondrial DNA.

Mitochondria are the organelles within every cell that produce energy from the foods we eat; they are the cell's power plants. Each mitochondrion contains enough DNA to code for about a dozen proteins—mostly enzymes—needed for energy production. Deterioration of these cell organelles has been linked with aging disorders such as late-onset diabetes, Parkinson's disease, and Alzheimer's disease.

Free radicals are not the only poisons produced by cellular metabolism. There are many others, and, if not discarded in a timely fashion, they will age and destroy our cells and ultimately our bodies. Regrettably, the cells' waste disposal systems are not perfect—toxic substances do accumulate over time. One such culprit is *lipofuscin*, a granular, fatty, yellowish-brown pigment that is a waste product of lipid metabolism. When it collects in cells of the skin it produces the "liver," or age, spots commonly seen in the elderly. When lipofuscin collects in brain cells and muscle cells, as it is wont to do, it clogs and kills the cells.

Volumes have been written on the subject of random cellular damage as a causative factor in aging. The above examples merely scratch the surface of the subject. What is less understood is why certain persons are so much more susceptible to these ravages of the body than others. Why will a Frenchwoman live for 117 years while her neighbor might age so rapidly that she will look old and wrinkled before she reaches her teens?

The answer lies in the body's ability to repair or reverse random damage to cells and tissues. Were it not for repair mechanisms, humans would not survive a day, let alone years. In the group of inherited disorders known collectively as *progeria*, one or another of these mechanisms is defective. The cause may be a missing enzyme or two that ordinarily wipe out free radicals. Perhaps a DNA or protein-repair enzyme is nonfunctional. Wherever the breakdown occurs, its consequences are disastrous. Aging proceeds at an incredibly rapid rate. Progeria sufferers will look like little old men or women—bald, wrinkled, bent over, arthritic—before they reach puberty. Death, usually from advanced cardiovascular disease, occurs by the early teens.

Clearly, longevity depends upon damage-control systems that maintain a high level of efficiency. These systems must monitor and regulate constituents of the blood, repair proteins and DNA damaged by the assault of radiation and chemicals (such as free radicals), heal wounded tissues, dispose of and replace dying cells, tirelessly do battle with invading microorganisms, and excrete toxins and cellular wastes. When these systems begin to fail—when rate of damage exceeds rate of repair—we age.

Hormonal Influences

Hormones are powerful substances produced by the glands of the endocrine system. They have a wide range of effects on different cells and tissues of the body. *Thyroxin*, the thyroid gland secretion, controls cellular metabolism—the rate at which cells utilize oxygen in energy-releasing reactions. In tadpoles, thyroxin initiates the animal's metamorphosis into an adult frog, acting, in effect, as a biological clock.

Hormonal clocks are common in humans as well. At puberty, the pituitary gland and the gonads release hormones that bring on sexual maturity. Later in life, the shutting down of these hormonal pumps in women results in menopause. Pattern baldness in men, yet another age-related event, is caused by the male hormone *testosterone*. As one might expect, castration prevents balding. I think I'll stick with my hairpiece.

What is the significance of all this? Clearly, hormones, or their absence, are involved in many physiological changes associated with aging. Might they not more directly control the actual aging process?

Indeed they might. In 1990, an experiment was performed that proved to be the biggest surprise in the history of aging research. Dr. Daniel Rudman, an endocrinologist, succeeded in turning back the aging clock by injecting a group of elderly men with human growth hormone. Growth hormone is a potent secretion of the pituitary gland that breaks down fat as it builds bones, muscles, and internal organs. It also strengthens the immune system and promotes healing of wounds. Unfortunately, by age sixty the pituitary stops making this vital antiaging hormone. Dr. Rudman administered growth hormone to twelve men in their sixties and seventies three times a day for six months. The results were astonishing: Subjects became leaner and stronger. Internal organs gained mass and functionality. Skin became thicker and more youthful. In six months, ten to twenty years of aging had been reversed. When treatment stopped, all gains were lost.

While the pituitary is shutting down production of growth hormone, it may at the same time begin pouring out a death hormone called *DECO*, an acronym that stands for *de*creasing *c*onsumption of *o*xygen, which is its most obvious effect on cellular metabolism. Experiments show that DECO works by blocking the action of thyroid hormones. In its absence, the bodily functions of old rats returned to juvenile levels, and their longevity increased dramatically.

The adrenal gland also puts out a hormone called

DHEA (dehydroepiandrosterone), which has many of the beneficial life-prolonging effects of growth hormone. As we age, the production of this powerful hormone begins to ebb. Unfortunately, scientists are still pretty much in the dark as to how these hormones prevent aging on a cellular level.

Genetic Programs

Want to live to a ripe old age? Pick good parents. That is exactly what Dr. Michael Rose, of the University of California at Irvine, did for a colony of red-eyed fruit flies. He bred only long-lived specimens with other long-lived specimens. After sixty generations, some flies were living almost six months—more than three times the lifespan of the average fruit fly. In terms of human lifespans, we're talking more than 225 years.

Rose attributes the increase in longevity to hundreds of changes in the fruit fly's genetic makeup. Mixing and matching of genes over sixty rounds of selective mating had caused an accumulation of genes that promote longevity. One significant gene change resulted in the production of a more active form of *superoxide dismutase* (SOD), a free-radical-fighting enzyme that is extremely important in cellular damage control.

Another favorite research subject of the gerontologists is a soil-dwelling nematode or roundworm the size of a comma. This fascinating creature matures in three days, defecates every fifty seconds, and dies, on cue, after twenty days. What causes this precision with which the nematode conducts its life? Gerontologists believe they have found a literal "death" gene, which, starting on day three, begins producing proteins that plunge the worm into a fatal decline. When this gene is inactivated, the worms can live twice as long as they otherwise would.

These findings should come as no great surprise. Ever since the elucidation of the structure and function of DNA almost forty years ago, scientists have felt that the ultimate control of aging processes must reside in the genes. Whichever enzymes or hormones are involved in age prevention, their synthesis is under the strict control of specific genes. It is only recently, however, that these specific longevity genes and the metabolic pathways they code for have been identified.

Much of the research to date has been with simple organisms—insects, worms, yeast, mice. They have much shorter lifespans and can be manipulated in ways not possible with human populations. (One cannot selectively breed humans for sixty generations.) But successful results with lower life-forms can pay handsome dividends in terms of human aging research, for the genes that promote long life in other organisms are structurally similar to longevity genes in humans. The DNA sequences of the two can, in fact, be virtually identical. Once these genes are identified and their sequences determined in nonhuman cells, gerontologists can search for their human counterparts. This is now being done for a number of antiaging genes, such as LAG-1 (longevity-assurance gene 1), recently identified in yeast cells.

The search for human genes that prevent aging employs a technique known as *tissue culture*—growing human cells outside the body. Working with these cells, scientists have already found a number of longevity genes. This is no small accomplishment when one considers that each human cell nucleus has nearly 100,000 different genes, made up of 3 billion smaller DNA subunits called base pairs, spread over 23 distinct and different chromosomes. (Only one chromosome of each pair in a cell is considered when mapping the totality of human genetic material; it is called the human genome.) A written record of the base-pair sequences of the human genome would require the equivalent of 200 telephone books of 1,000 pages each. Nonetheless, genes that promote aging in skin, blood, and brain cells have been located or mapped. On chromosome six are a cluster of longevity genes that code for SOD as well as a DNA-repair enzyme and a substance that carries hormonal messages into the cell. Genes involved in aging have also been discovered on chromosomes one and four.

As far back as 1965, Dr. Leonard Hayflick reported that certain human connective tissue cells in culture would divide about fifty times and then stop, dying soon afterward. Other types of human cells showed this same remarkable property. Placenta and foreskin cells taken from embryos or newborns divided sixty to eighty times and then failed to divide any more. Replication of the cells also slowed down as they aged.

What was most impressive was that these cells seemed to remember how many times they had divided. Cells that were frozen after twenty generations of growth and then thawed years later would divide only an additional forty to sixty times. And cells taken from older people replicated a fewer number of generations before death than those taken from newborns.

These remarkable findings suggest the presence of a genetic aging clock within cells. Its significance to an understanding of aging is pointed out by Professor Arnold J. Levine, of Princeton University: "Some scientists believe that these experiments are measuring an inherent property of the aging process. They hypothesize that all cells are programmed to have a limited life span of sixty to eighty generations and that we use up these divisions with age. Aging would then be defined as a declining ability to reproduce our blood cells, skin cells, bones, and so forth."

Geneticists began looking for this cellular clock of

aging in human cells. What they found were *telomeres*—strings of repeating DNA segments that dangle from the ends of chromosomes. Whenever a cell divides, it loses a number of telomere segments. As the telomeres shorten, the cell ages and eventually dies.

The discovery of this property of telomeres raises more questions than it answers. Are they merely clocks that keep track of a cell's declining ability to replicate, or do they play an active role in determining the fate of the cell? Evidence points to the latter. Cancer cells, which never stop dividing and, consequently, never die, have an enzyme, telomerase, which constantly synthesizes new telomere segments.

The Future

How far have we really come? Are gerontologists any better off today in their understanding of aging than the blind men and their elephant? Yes and no. Research has uncovered a number of cellular mechanisms—enzymatic and hormonal—that retard as well as accelerate aging. Many more are yet to be discovered. We are even pinpointing the bits of DNA, the genes, that code for these biochemical pathways. But if we are to conquer aging, we must learn what turns on the genes themselves. The key to unlocking the secret of eternal youth lies in a knowledge of why certain genes flip on at precise times while others flip off. It is these genetic switches that ultimately control aging. Unfortunately, how they operate is still very much a mystery.

What lies ahead? That depends very much on whom you speak to. Dr. S. Michal Jazwinski, of Louisiana State Medical Center, the foremost authority on aging in yeast, is quite optimistic: "Possibly in thirty years we will have in hand the major genes that determine longevity and will be in position to double, triple, even quadruple our maximum life span of

120 years. It's possible that some people alive now may still be alive 400 years from now." Talk about beating the social security system!

A more conservative view is expressed by Dr. Anna McCormick, a molecular biologist at the National Institute of Aging. "To stop aging you have to fine-tune the actions of maybe more than a hundred genes, many of which perform more than one function. One wrong step and you could wind up with a very sick organism." Dr. Vincent Cristofalo, of the University of Pennsylvania, concurs. "No magic bullet is going to put an end to aging. It's too complex. We may progressively stop certain aspects of aging over the next hundred years. We definitely won't stop all of aging in the near future."

In the meantime, we continue to search for the Fountain.

What's the Big Deal with Plastics?

"I have just one word to say to you. Just one word: plastics."

An enterprising uncle uttered this bit of advice to Benjamin Braddock, played by a young Dustin Hoffman, in *The Graduate*. Although this remark was meant to be humorous at the time the movie was released (1967), it turns out that since that movie hit the silver screen it was not bad advice. Plastics have completely reshaped our civilization. If everything on Earth were to disappear except plastics, the world as we know it would still be quite recognizable. No other material—not wood, not metal—is more ubiquitous. Plastics are molded into toys, clocks, radios, televisions, jewelry, automobile parts, soda containers, and trash cans. They are woven into clothing. We paint and insulate our homes with plastics. Formica, used in furniture construction, is a type of plastic; so are the epoxy glues and adhesives in common usage. The rubber tires your car runs on, the Teflon coating your pots and pans, the waterproofing you spray on your boots, and possibly the surgically replaced valve in your heart—all are made of plastic. *The Graduate*, like all motion pictures, was recorded on cellulose acetate, a type of thin plastic film.

Synthesis of Plastics

But what are these extraordinarily versatile, man-made substances we call plastics? How are they created? These are not easily answered questions, for plastics include an extremely diverse group of substances. There are, however, certain features that are common to all plastics and that, in fact, help define them.

Plastics are organic compounds. This means that the main part, the backbone, of the plastic molecule is a chain of carbon atoms. (An exception are the silicones, which have a molecular chain of alternating silicon and oxygen atoms.) In most cases the chain is quite long and consists of smaller, repeating subunits called "monomers." The monomers are chemically bonded to one another, like many pearls on a necklace. The process of joining the subunits together is called *polymerization*, and the resultant molecular chain is a *polymer*. Polyethylene, used to make plastic wraps and garbage bags as well as a host of molded products such as toys, kitchenware, and bottles, is a long-chain molecule of many ethylene monomers.

ethylene monomers

```
H   H   H   H   H   H   H   H
|   |   |   |   |   |   |   |
C — C — C — C — C — C — C — C ...
|   |   |   |   |   |   |   |
H   H   H   H   H   H   H   H
```

polyethylene molecule

Polyvinyl chloride (PVC), a close relative of polyethylene, is a polymer of the vinyl chloride monomer.

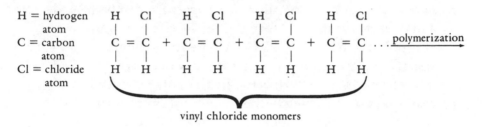

H = hydrogen atom
C = carbon atom
Cl = chloride atom

vinyl chloride monomers

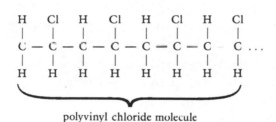

polyvinyl chloride molecule

Being both cheap and versatile, PVC is probably the world's most widely used and industrially important plastic. Pipes and pipe fittings, computer floppy disks, garden hoses, building sidings, wire and cable insulation, food packaging—these are just a few examples of its many applications.

Some plastics are made of thousands of monomers linked together into truly huge molecules. All the monomers in a plastic molecule need not be identical. ABS, a popular synthetic rubber, is made by polymerizing three different monomers. The linking of different monomers into new carbon chains is what the science of plastics is all about. Between 1938 and 1941 the German chemical company Farben was creating new polymers at the rate of one a day. Today there are tens of thousands of different plastics with a fantastic range of properties.

The word *plastic* comes from the Greek *plastikos*, meaning "fit for molding." Its name describes a key feature of all plastics: when heated they become soft and pliable. *Thermosoftening* plastics, also called *thermoplastics*, can be heated and cooled repeatedly. Most plastics—celluloid, polyethylene, polystyrene, nylon, PVC—are thermosoftening. *Thermosetting* plastics, on the other hand, can be softened and molded only once—during the initial polymerization process. Additional heating will only cause them to turn brittle and become harder. Thermosetting plastics include Bakelite and other phenol-derived plastics, formica, plastic glues such as the epoxies, and polyurethanes used in paints and varnishes.

A closer look at the molecular structure of plastics will explain why they behave as they do. The thermoplastics are a mass of coiled, entangled polymer molecules. These polymer chains lie very close to one another—close enough for there to be *cohesion*, or molecular attraction between them. But the individual molecules can still move freely over one another, giving the substance plasticity. Heating the plastic causes molecular vibration. The polymer chains separate a bit and can more readily slide over one another. The plastic becomes more fluid, or pliable.

With the thermosetters, things are radically different. Heating causes cross-link chemical bonds to form between these entangled polymer chains. The individual molecules lose their freedom to move along one another and the plastic becomes thoroughly rigid and brittle. In effect, the plastic mass has become one large molecule.

Large, carbon-based molecules are not something new. Long before scientists were creating them in plastics laboratories, living cells were synthesizing them. They are the organic molecules—sugars, starches, lipids, proteins, nucleic acids—that are the fabric of life itself. Small wonder, then, that many natural resins were the forerunners of today's plastics. *Amber*, resin that dripped from prehistoric pine trees and entrapped many now-extinct insects, is still used today to make jewelry. Natural rubber was used as far back as the eleventh century by the Maya Indians of Central and South America. It is made from a milky liquid tapped from under the bark of certain trees. The first transatlantic telegraph cables—between England and America—were insulated with *gutta-percha*, a rubberlike sap of the sapodilla tree.

The first plastics, in fact, were produced by experimenting with and chemically altering naturally occurring organic compounds such as cellulose, a woody plant fiber, and casein, the protein that can be curdled out of milk. Soya beans, potatoes, corncobs, coffee beans, peanuts—all at one time or another were grist for the plastics mill. The story of plastics is a fascinating one and, strangely enough, it all began with the game of billiards.

History

By the 1860s billiards had become very popular in the United States. The game was played with balls made from the ivory of elephant tusks. So great was the demand for billiard balls that elephant herds were decimated and ivory, coming in from Africa, became very scarce. The situation was such that in 1863 a New York billiards company offered a $10,000 prize to anyone who could produce a satisfactory substitute for the ivory ball.

This generous offer caught the attention of a young Albany printer named John Wesley Hyatt. Hyatt was not a trained chemist, so he followed the lead of experimenters in England by mixing together with adhesives such plant products as sawdust, shredded paper, and chopped-up cotton rags. The resultant concoction was a cellulose pulp. Cellulose, a plant fiber mentioned earlier, is a naturally occurring polymer that makes up the outer walls of plant cells. It constitutes the indigestible roughage fibers of the fruits and vegetables we eat.

Hyatt's mixture was easily molded into balls. When hardened, however, the balls lost their roundness, cracked, or shriveled up. No mixture of substances seemed to do the trick. Then, in desperation, Hyatt added nitric acid to the cellulose mixture. What he produced was *cellulose nitrate*.

Little did John Hyatt realize how dangerous his billiard-ball experiments were. With enough heat and pressure or with slightly different ingredients, the cellulose nitrate could have easily produced guncotton and exploded in his face.

Hyatt's next step was to mix the cellulose nitrate with camphor. The result was magical: a product was formed that could be molded into any shape when heated. Upon cooling, it became so hard and strong that it could be planed or worked on a lathe, like wood.

Hyatt called his discovery *celluloid*; it was a huge commercial success. The age of man-made plastics had begun. Everything from automobile window curtains to imitation collars and cuffs to jewelry to false teeth were made of celluloid—more than 2,500 different articles in all. Hyatt's

Celluloid Manufacturing Company produced the first motion picture films in 1882.

But Hyatt never won his $10,000 prize. Celluloid proved too brittle for billiard balls. It was also a highly flammable substance that produced toxic fumes as it burned. Anyone wearing a celluloid shirt collar or using a celluloid comb while smoking a cigarette stood the very real risk of going up in flames. Hold a lighted match to a Ping-Pong ball, which is still made of celluloid, and the danger becomes apparent.

The next great advance in the development of plastics was another forty years in coming. And it had nothing to do with finding an ivory-substitute billiard ball, although a satisfactory one had yet to be found. The push now was for a synthetic shellac. Made from a natural resin produced by tiny insects in India, shellac proved to be an excellent electric insulator. Unfortunately, it took 150,000 bugs six months to produce enough resin for one pound of shellac. As a result, the rapidly expanding electrical industry put enormous demand on a very limited shellac supply.

Enter Dr. Leo Baekeland, an accomplished chemist who had already invented a type of photographic paper and made a small fortune selling it to Eastman Kodak. Baekeland began mixing two different organic chemicals—*phenol* (carbolic acid), a disinfectant distilled from coal tar, and *formaldehyde*, the acrid-smelling preservative known to all biology students. What he got was a viscous, syrupy substance that he could not pour out of his test tube. To Baekeland's amazement, heating the material did not soften it—it made it rock hard.

Baekeland had produced a long-chain polymer of the phenol and formaldehyde molecules. He humbly named it *Bakelite*, in honor of himself. The suffix *ite*, when referring to plastic, means that it is hard and not flexible or rubbery. Bakelite was the first completely synthetic polymer, produced by true polymerization of smaller organic molecules and not through alteration of preexisting natural polymers, as in the case of celluloid.

Dr. Baekeland's discovery was probably the single greatest achievement in the history of plastics research and development. Mixed with other substances such as asbestos fibers, it went on to become the workhorse of the plastics industry until World War II. A material that was both heat-resistant and an excellent insulator, Bakelite found ready application in toasters, pot handles, electric switches, plugs and receptacles, radio chassis and parts, car distributor caps, and the like. In September 1924 Baekeland appeared on the cover of *Time* magazine.

But the story of plastics does not end with Baekeland. In 1928 the U.S. chemical firm E.I. du Pont de Nemours, set up a laboratory headed by Wallace Carothers to study polymer chemistry. Ten years later, they announced creation of a miraculous plastic with amazing elasticity and tensile strength—*nylon*. In May 1940 the first stockings knit with nylon fibers went on sale. Nylon and other plastics such as rayon, dacron, orlon, and the polyesters were to change the face of textile production.

When World War II came along, plastics greatly aided the war effort. Cut off from the Japanese silkworms, America turned to its newly developed nylon to make tents, parachutes, and parachute lines. Such vital equipment used up half of all the nylon produced at that time. Women were even encouraged to donate their nylon hosiery to the war effort. Betty Grable auctioned her nylon hose for $40,000 at a war-bond rally.

After the bombing of Pearl Harbor, America's sources of natural rubber in the Pacific were jeopardized. Rubber was essential for truck, jeep, and airplane tires. Luckily, DuPont and Farben had discovered synthetic rubbers in the early 1930s. To solve the crisis, President Franklin D. Roosevelt created the Office of Rubber in 1942. Fifty synthetics factories were built throughout the country, and synthetic rubber production rose from 20,000 tons in 1942 to 700,000 tons in 1944.

The end of World War II saw an unparalleled flourishing of the plastics industry. Polymers such as *Teflon*—immensely important during the war as insulation for radar equipment and for work on the Manhattan Project, which developed the first atomic bomb—were put to peacetime use. Organic chemists began a systematic study of the polymerization process. What was once largely a hit-or-miss operation of mixing and heating monomers became a very exact science. Today polymer chemists know the precise structures of the complex molecules they are synthesizing. They can accurately predict the outcome of untried experiments and thus tailor-make plastics with particular properties for particular needs.

Polyimides, for example, because of their extreme resistance to temperature, chemicals, and mechanical wear, are being synthesized for use in space shuttles and particle accelerators. The U.S. Navy has produced a silicone membrane for underwater breathing that filters out the oxygen dissolved in water. Epoxies exist that can seal deep and possibly fatal wounds. The nose cone of the Concorde is made of a glass-reinforced plastic. Clear bulletproof windows and windshields are made of a tough transparent *polycarbonate* plastic. A recently developed plastic has the remarkable property of rapidly sealing itself when cut, making it especially useful in the construction of gas pipes.

Indeed, plastics have had a glorious past. And their future looks just as promising. The ways in which monomers can combine to produce different products seem endless. Now polymer chemists are even rearranging the polymer chains that make up a plastic. Instead of existing as a random mass of entangled molecules, they can be lined up next to one another. Faye Flam, Washington correspondent for *Chemical Weekly*, explains, "One chemist compares the process to 'uncooking spaghetti,' because scientists take the coiled, spaghetti-like polymers that make up plastic, straighten them out, and put them back together in a parallel fashion—something like the way spaghetti comes in the box."

What one gets by doing this is an *oriented plastic*. Oriented polymers are the wave of the plastics future. They can make plastics ten times stronger, stiffer than steel, and more electrically conductive than copper. The first fruit of this discovery was a product known as *Kevlar*, which has been used to make bulletproof jackets and protective clothing for chain-saw operators.

The potential of oriented plastics is enormous. Aircraft could be protected from static electricity and lightning and be made invisible to radar. A transparent, electrically conductive polymer could be used in converting sunlight to electricity in solar-energy collectors. The semiconductive nature of some oriented polymers suggests their possible use as transistors. Polymer expert Alan Heegar, of the University of California at Santa Barbara, compares the prospects for today's new polymers to the bright future of ordinary plastics when Dustin Hoffman was getting his bit of avuncular advice. "Our work has the potential to start a revolution like the first one," he states. "We just have to see how far it will go."

I have just two words to say to you. Just two words: oriented plastics.

Viruses— the Dinner Guests That Stayed

One of the worst plagues ever to afflict humankind occurred in 1918. It made half the world's population sick as it swept across the planet and killed millions of people— estimates go as high as 40 million—in a matter of months. So desperate did the situation become that the Commissioner of Public Health in Chicago told police to arrest anyone sneezing in public. San Francisco passed a law forcing people to wear surgical masks over their mouths and noses in public. Violators were arrested as "mask slackers."

This dread killer was influenza, commonly called the flu. Other less-serious flu epidemics occur periodically. They are caused by the tiniest of all living things, the *virus*. Many of the most terrible scourges of the past—smallpox, yellow fever, polio—as well as milder infections such as the common cold are of viral origin. So are 20 percent of all cancers. AIDS is caused by a virus originating in Africa, and many virologists believe there are a host of other very deadly viruses lurking in Asia, Africa, and South America, waiting to emerge.

Sound scary? It is. In 1989, several research workers in Virginia became ill while working with monkeys imported from the Philippines. A virus was isolated that looked identical to the *Ebola* virus. Before the discovery of AIDS, Ebola was the deadliest human virus known; that is, it had the highest mortality rate of infected persons. It causes a hemorrhagic fever accompanied by vomiting, internal bleeding, shock, and death.

Until that time, Ebola had never turned up anywhere outside of Africa. Mercifully, the Philippine-imported virus proved to be similar, but not identical genetically, to Ebola, and it was far less deadly. There is still, however, great cause for concern. The ease with which people travel from continent to continent today means that new viral pathogens can spread much more rapidly. Changing climates due to global warming are creating new, hospitable environments for animals that are viral carriers. It is estimated that about one hundred *arboviruses*—those transmitted through the bites of mosquitoes and ticks—cause disease in humans, and at least twenty of these are capable of becoming epidemic.

Joshua Lederberg, 1958 Nobel Prize winner in medicine and physiology, expressed this concern in December of 1990. "It is still not comprehended widely that AIDS is a natural, almost predictable, phenomenon. It is not going to be a unique event. . . . There will be more surprises, because our fertile imagination does not begin to match all the tricks that nature can play." What sort of tricks was he talking about? Perhaps it is time to take a closer look at the virus.

Discovery

Although viruses have probably been around since cellular life began, they were discovered only a scant hundred years ago. In 1892, a Russian scientist, Dmitri Ivanovski, was investigating tobacco mosaic disease, which causes mottling of tobacco leaves. He passed the sap of infected plants through a porcelain filter that was believed to trap all types of microorganisms, including bacteria, the smallest known pathogens. Surprisingly, the filtered liquid still caused infection. Could it be the toxin of a germ that passed through the filter? Not likely, since the sap could transmit the fully virulent disease successively through any number of plants. This nondilutable property suggested an entity that was reproducing itself in the plant.

Soon other diseases were found to be caused by this "soluble living germ," which came to be called a virus (from the Latin word meaning "poison"). Foot-and-mouth disease of cattle was the first animal illness shown to be transmitted by a filterable agent smaller than any known bacterium. By 1900 a human disease, yellow fever, was proven to be of viral origin. Even bacteria were found to be infected by these filterable agents. Indeed, viruses infect everything that is alive. Viruses that attack bacteria are called *bacterio-phages*; these were the first ones to be extensively studied. The advent of tissue culture (growing animal and plant cells in a dish) made widespread research into animal and plant viruses possible. Since the invention of the electron microscope in the 1930s, we can actually see viruses.

What has all this investigating taught us? To begin with, viruses differ from all other living things in that they are not cellular. All the structures within a cell necessary to perform the life activities of eating, energy production, growth, and response to environmental change are absent in a virus. It is, in fact, nothing more than a tiny, lifeless, totally inert particle—as long as it remains outside an actual living cell. Tragically, it is made to get inside the cell.

Description

All viruses consist of two parts: a nucleic acid core and a protein coat surrounding the core. In some cases, there is an additional fatty, or lipid, envelope. It is the function of the protein coat and lipid envelope (if present) to attach the viral particle to a cell membrane and—somehow—get the virus into the cell. This is not easy. The surface of the viral coat must fit exactly into "receptor" sites on the cell membrane. If the fit is not precise, then attachment and subsequent penetration into the cell cannot occur. Even in an ideal matchup, probably only one out of every few thousand collisions between virus and suitable cell results in proper binding of the two. The exactness of fit necessary for viral binding or attachment explains why viruses are usually species-specific—they will not infect cells of totally different species. Notable exceptions are the rabies and influenza viruses, both of which have a wide range of hosts.

Not only will viruses seldom infect totally different species, but quite often they are specific to particular types of cells within an organism. The *hepatitis B* virus targets liver cells. *HIV* goes for particular binding sites, or markers, on T4 white blood cells.

Once the virus is attached to the cell, there are several ways that it can penetrate the membrane and enter the cell. It may cause the cell membrane to infold and pinch off a tiny vesicle with the virus inside. Viruses with fatty envelopes may fuse their envelopes with the cell membrane, penetrating it and letting the rest of the virus into the cell.

When a virus does get inside the cell, it must make

more particles like itself. To understand how this is done, we must learn more about the nucleic acid core of the virus.

Nucleic acids are found in every living cell as well as in viruses. They are of two basic types—DNA and RNA. DNA is the stuff genes are made of. This means that a particular DNA has stored within it the information needed to construct and maintain a particular organism. Trees are different from humans because tree DNA (found in every tree cell) is different from human DNA (found in every human cell). And no two humans are ever exactly alike because their DNAs are slightly different.

How does this DNA determine the form and function of an organism? Very simple. It controls the proteins that the cells of an organism produce. There are more than 100,000 different kinds of proteins found in living things. Some are structural components of cells while others, called *enzymes*, control chemical reactions in the cell. Different DNAs make different proteins. Different proteins make different types of organisms.

So far so good. But where does RNA fit into the picture? The principle function of RNA in cellular chemistry is to act as a messenger. It takes information from the DNA (found in the nuclei of all cells) and delivers it to the sites of protein synthesis (found outside the nuclei). The DNA acts as a template to create the correct RNA. The RNA, in turn, directs the building of the proper proteins.

Now we return to our friend the obligate intracellular parasite, better known as the virus. Viruses have either RNA or DNA cores but never both. And in viruses, the RNA is not a messenger molecule but the real thing—the genetic material. The smallest viruses, such as those that produce the common cold, have enough nucleic acid to code for only several different kinds of proteins. By contrast, the largest poxviruses have enough DNA to make several hundred proteins.

Whatever the amount of their nucleic acids, all viruses must accomplish the same thing: they must shut down the production of cellular materials and force the cell to make more viral particles. The usual sequence of events is as follows:

1. Proteins are produced that are predominantly enzymatic in nature. They catalyze the production of many thousands of copies of viral nucleic acid.
2. After viral nucleic acid is synthesized, the structural-coat proteins are produced.
3. The virus is assembled by having the coat protein form as a shell around the nucleic acid core.
4. These new viral particles are released, sometimes—but not always—killing the cell in the process.

The way in which a virus can slip into a cell and coerce it into producing more viral particles exactly like itself is an incredible bit of piracy. To pull it off, a virus needs the correct combination of precise proteins and nucleic acids. Such a unique assembly of proper molecular components is not accidental or coincidental. It comes about through a very long association in which viruses have adapted and are continually adapting to their hosts. In fact, virologists now believe that viruses and cells must have existed together and *coevolved* since the beginning of life on this planet.

One theory proposes that viruses, because they are simpler than cells, actually predated them in origin. According to this hypothesis, nucleic acids increased in complexity until they became the stuff of cells. Along the way, simpler DNA or RNA strands were left behind to learn the ways of parasitism—to become the stuff of viruses. Although once popular, this theory is now considered unlikely. A more probable scenario is that viruses have evolved from bits of cellular genetic material that escaped from their cells eons ago.

Over time, according to this "escaped gene" hypothesis, genes developed the ability to be independent, self-replicating, intracellular parasites—viruses. The more successfully adapted the viruses became to their host cells, the less damage they inflicted. That is the pattern of viral evolution, and there is even evidence that the AIDS virus is becoming slightly less virulent. Remember—if a virus kills its host, it is, in effect, cutting off the hand that feeds it. Most viral infections, in fact, do no real damage at all, going totally unnoticed. But oh, what damage, what pain and suffering those imperfectly evolved viruses cause!

Diseases

Viral infections are basically of two types, acute and persistent. An acute viral infection is one in which the virus vigorously invades a tissue or tissues of the body. There is rapid multiplication of viral particles and an equally rapid immune response. If things go well, the virus is defeated and a complete cure is effected. If not, severe symptoms and death may result.

Acute infection is virus-host interaction at its simplest, and it is a common course for many viruses to take. Measles, mumps, meningitis, rabies, the common cold, and influenza are diseases in which infection is acute and the conflict lively between virus and immune system. An aggressive immune response is what makes *vaccination* against many acute diseases possible. A vaccine contains virus that has been killed or made weak by heat or chemicals. Its similarity to the real virus tricks the immune system into a vigorous response, which can last a lifetime. Vaccination has been remarkably successful in wiping out acute viral killers such as smallpox and poliomyelitis. More than 60 million people died of smallpox in Europe between 1650 and 1750. Thanks to vaccination there has not been a single case of smallpox in the world since October 1977.

When a viral infection is acute, there is a head-on collision between virus and immune system. Recovery from illness means complete elimination of the virus. This, however, is not always the case. In fact, we are discovering that it is not usually the case. Many, if not most, viruses are able to retreat during battle and literally hide from the immune system. In this way, they can linger in the body for many years. These are called *latent* viruses, and they produce persistent viral infections.

The precise nature of latency is poorly understood; viruses that exhibit this property are difficult to deal with and cause a wide range of serious diseases. Among the *herpesviruses*—masters of latency—this ability to "hit and hide" causes recurrent or chronic infections. The word *herpes* comes from the Greek verb "to creep" and aptly describes the spreading lesions of *herpes simplex 1*. These are the typical cold sores in and around the mouth area. Genital herpes is caused by a closely related virus, *herpes simplex 2*. When the herpes virus is not actively multiplying in skin and mucous membrane cells, causing lesions, it retreats within certain nerve cells. These quiescent periods can last months or even years.

Chicken pox, also caused by a herpesvirus, follows a similar course of action. After initial infection, the virus becomes latent, lying low within the nerve cells of the vertebrae. In most people, the virus remains dormant for life. It can, however, reactivate, causing the painful back rash characteristic of the disease called shingles.

Epstein-Barr is yet another herpesvirus. When actively

replicating in B white blood cells (those that produce antibodies) it produces the acute infection mononucleosis. It can, however, reside for many years within B cells—the latent period—resulting in chronic fatigue syndrome and possibly certain cancers.

A number of different viruses have been linked with cancer. Hepatitis B (serum hepatitis), a viral infection, is the leading cause of liver cancer. But it is the retroviruses, so named because of a reverse (retro) step in the replication of their nucleic acid, that have raised cancer production to an art form. They accomplish this by a phenomenon known as *integration*—the incorporation of viral DNA into the host-cell genome. The virus, in effect, becomes part of the cell, buried safely in the nucleus, far from the immune system. This is latency at its best. Unfortunately, integration adversely alters the genetic makeup of the host cell and can, over time, cause it to become cancerous.

In Los Angeles during 1980 and 1981, five cases of a rare pneumonia were reported to the Centers for Disease Control in Atlanta. By 1982, New York was reporting cases of a rare cancer, Kaposi's sarcoma. The sudden frequency of these diseases signaled the beginning of the AIDS pandemic. It is common knowledge that AIDS is caused by the human immunodeficiency virus (HIV). What may not be so widely known is that HIV is a retrovirus, with the ability to be latent. It specifically attaches to and infects white blood cells called *helper T*, or *T4*, cells. These cells produce chemicals essential to the proper functioning of the immune system. Once inside the T4 cells, this virus can lurk unnoticed for a dozen or more years, hidden inside the very cells meant to destroy it. A person infected with the virus in this manner is said to be HIV positive.

HIV is passed from cell to cell as infected cells make contact and fuse with healthy ones. This unusual mode of transmission protects the virus from exposure to antibodies in the blood. Then something—most likely another unrelated infection—triggers the rapid multiplication of HIV. It now begins killing T4 cells in earnest, marking the onset of full-blown AIDS.

The variety of ways in which viruses can affect disease is both remarkable and terrifying. They are even capable of turning the body's own immune system against itself, producing what are termed autoimmune diseases. In many cases, autoimmunity may be a natural consequence of infection. During the course of penetration and multiplication, viruses leave bits of themselves embedded in the host cell's membrane. Vigilant white blood cells will recognize these cells as foreign and destroy them. This destruction of infected cells is vital to ridding the body of the virus, but sometimes the cure is worse than the disease. During hepatitis B infection, it is believed that more liver damage is caused by the immune system's attack on liver cells than by the actual virus.

Autoimmunity may also be induced accidentally, as a result of viral infection. In a phenomenon known as "mimicry," the outer coat of the virus may be similar enough to the surfaces of certain host cells to cause white blood cells to target friend instead of foe. It is now believed that autoimmune diseases such as rheumatoid arthritis, lupus erythematosus, and multiple sclerosis may have viral origins.

Ann Giudner Fettner, in her book *Viruses: Agents of Change*, states that "today is the day of the virus." This is an understatement. Nothing being studied in medicine, nothing in biology, is more important.

In 1985 researchers completely mapped the chemical structure of a common-cold virus and created a three-dimensional model of the viral particle. It required the use of an atom smasher to take x-ray photographs of the virus's atoms and a supercomputer to process the photographic information.

This was a remarkable feat of technology, yet our present knowledge of viruses as pathogens is tragically incomplete. The diseases presently known to be caused by viral agents are only the tip of the iceberg. They may be causative factors in diseases as diverse as severe depression, diabetes, schizophrenia, coronary heart disease, and who knows what else. There are more than three hundred varieties of rhinovirus, all causing the common cold. Wild animals often act as reservoirs for any number of human viruses, making them impossible to eradicate, even with universal vaccination. A new flu virus appears every few years in humans after hanging out in pigs or ducks. During that time it has mutated (altered its nucleic acid) and acquired a new protein coat. HIV is constantly mutating; in 1986 a new strain, HIV-2, was identified. Others are sure to follow. There is even a growing body of evidence that the flu virus of 1918 was the cause of a severe form of catatonia in survivors of that dread epidemic. The disease, called *encephalitis lethargica* (sleeping sickness), was the subject of Oliver Sachs's bestselling book *Awakenings* (later made into a motion picture starring Robert DeNiro and Robin Williams).

The virus has been with us for several billion years, and it has made itself very comfortable—the guest that came to dinner and would not leave. To learn about viruses is to learn about the creation and evolution of life. Of all the DNA in the human cell, only 1 percent appear to have any function in coding for proteins. The other 99 percent are probably a fossil legacy of viral parasitism—genetic material left behind by viral infections both ancient and contemporary.

Like it or not, viruses are here to stay. We might even be wise to enlist their special talents to cure some of the 4,000 known human genetic disorders. What better way of getting the proper gene into a defective cell than by hooking it up to a retrovirus that has been rendered noninfectious. This sort of gene therapy also might be used to correct the DNA of cells that have turned cancerous.

In many ways, our present understanding of viruses is comparable to our knowledge of bacteria thirty or forty years ago. There is so much still to be discovered. In the meantime, let us hope that those "surprises" nature has in store for us will not be too devastating.

Can People Be Put into Cold Storage?

Do not go gentle into that good night,
Old age should burn and rave at close of day;
Rage, rage against the dying of the light.

These words were the entreaties of poet Dylan Thomas to his gravely ill father. Were he a cryonicist, he might have offered the following bit of advice: "Freeze, wait, reanimate." It is a favorite war cry of *cryonics* researchers: scientists who believe that the body can be frozen and thawed back to life at a future date, when medicine is advanced enough to cure the illnesses and aging processes that cause premature death today. Once relegated to the realm of science fiction, cryonics is steadily gaining respect within the scientific community.

Cryonics should not be confused with *cryogenics* or *cryobiology*, two other cold-temperature sciences. Cryogenics deals with the production of temperatures approaching absolute zero and the study of the behavior of matter at these temperatures. The discovery of superconductivity and superfluidity—very low-temperature phenomena—are two of the success stories of cryogenics. Cryobiology, a recognized branch of medical research, studies the effects of extremely low temperatures on living cells and tissues. Frozen skin, blood, corneas, bone marrow, sperm, ova, and even embryos are regularly stored for future use and thawed successfully. Unfortunately, organs such as hearts and livers do not fare as well. Presently they can be kept, frozen, for a day or so, after which they are no longer transplantable.

Such difficulty in organ preservation makes cryobiologists very skeptical as to the possible success of cryonicists. They disparagingly refer to them as "body freezers," and the Society for Cryobiology forbids members from cooperating with cryonicists.

All this notwithstanding, cryonics is experiencing an upsurge of popularity. In 1968, the first human was cryonically suspended. By 1989, twenty people had allowed themselves to be put into a frozen state of suspended animation. It is estimated that membership in cryonics organizations (there are several worldwide) doubled in the last five years. According to the December 1992 issue of *Cryonics*, the Alcor Life Extension Foundation of California (largest of all the cryosuspension facilities) alone has twenty-five cryopreserved patients, with several hundred others signed up and waiting. In one month alone, they suspended three patients. Clearly, people want to cheat death, but how realistic are their hopes of success?

Problems

According to Dr. Paul Segall, scientific director of Trans Time, a cryonics group, "freezing is nature's way of saying 'time out.' . . . It's a way of putting life on hold." Of this there is little doubt. In cryopreservation, the patient is suspended in a tank of liquid nitrogen, which lowers body temperature to −320° F (−196° C). At this extreme temperature, cell degeneration—and it is this deterioration of the body's cells that must be prevented—that would normally take one second is slowed to *thirty trillion years*. Everything stops, literally frozen in time.

Frogs routinely survive Canadian and northern United States winters by freezing up to half the water in their bodies. When spring arrives, they thaw out and go about their lives. So what seems to be the difficulty with putting humans in cold storage?

Very simply, freezing damages the tissues of the body. As cells freeze, water seeps out of them. Additionally, there is naturally occurring intercellular fluid, a watery mix, which bathes all cells. When this water, clinging to the outer surfaces of cells, freezes, it produces ice crystals, which tear and puncture the cells. Such damage, when it occurs, is both inevitable and irreparable.

There are ways of reducing, if not totally eliminating, freeze damage. Freezing the body very rapidly is one of them. Ice crystals take time to form, and fast freezing prevents buildup of large, damaging crystals. This can be done with single cells and thin sections of tissue. Unfortunately, it is impossible to "flash" freeze all the internal tissues and organs of an entire human being. Considerable freeze damage must necessarily result. Too rapid a temperature drop also causes "shell freezing," a condition in which the outer tissues freeze first and are subsequently cracked by the expansion of inner tissues as they freeze.

So how do frogs do it? In much the same way antifreeze protects your car's engine from being damaged by freezing water. Nature has provided this amphibious creature with an antifreeze of sorts, a "cryoprotectant" called *glycerol*. Although it allows for freezing of much of the frog's watery fluids, the freezing occurs at *reduced temperatures* and there is virtually no cell or tissue destruction.

Armed with such knowledge, cryonicists have experimented with glycerol and a variety of other cryoprotectants, including car antifreeze itself (ethylene glycol). For example, dogs were cooled to several degrees *above* freezing after their blood was replaced with cryoprotectant. After four hours in a lifeless state (no heartbeat, no breathing) the blood was restored and the dogs were revived, apparently no worse for the wear.

While experimenting with different cryoprotectant blends, an interesting and important discovery was made. At certain concentrations, cryoprotectant "cocktails" do not freeze, no matter how cold they get. Instead, they become increasingly thick, or viscous, until they reach a solid, glass-like state. Researchers had stumbled upon the process of *vitrification*—solidification without freezing.

No freezing means no ice crystals and no cellular freeze damage—a major breakthrough in cryosuspension. Vitrified human embryos have been thawed and successfully implanted into surrogate mothers. The resultant progeny seemed healthy in every way. The process has even become somewhat of a cottage industry among dairy farmers: embryos from the best milk-producing cows are now vitrified for future implantation into surrogates. These achievements may allow for the future development of organ banks where vitrified, transplantable organs may be stored for long periods of time.

Vitrification, however, is not without its problems. In the thawing process, a temperature is reached where ice can form. This presents a serious problem to large cryopreserved organisms, such as adult humans, in which thawing cannot be accomplished with uniform rapidity throughout the internal tissues. According to Greg Fahy, Ph.D., a cryobiologist at the American Red Cross transplantation laboratory and discoverer of the vitrification process, "it is unlikely that vitrification will work on an entire body."

So where does that leave cryonicists? No matter what procedures they employ, present technology is unable to prevent at least some cell and tissue damage. In other words, cryopreserved humans cannot be successfully revived today. Thankfully, however, perfect cryopreservation today is not essential for successful resuscitation in the future. Resuscitation will simply have to wait for the technology to be developed to repair damaged cells. One such promise for future repair and revival of frozen bodies is *nanotechnology*.

The Future

The prefix *nano* means one billionth. A nanometer is one billionth of a meter, a unit of length small enough to measure the size of individual molecules. Nanotechnology refers to an emerging science that attempts to alter and rearrange individual molecules with control and precision. It is the science of dealing with and manipulating the ultra-ultra-small.

Natural systems have been employing nanotechnology for millennia. A peek into the living cell provides ample proof of this. Ribosomes—structures, or organelles, within individual cells—are *nanomachines* that build proteins from the proper amino acid molecules. Indeed, every cell organelle—nucleus, endoplasmic reticulum, golgi apparatus—is a *nanosystem* of sorts, recognizing, manipulating, and creating on a molecule-by-molecule basis.

The more researchers have learned about natural nanosystems, the more confident they have become that man-made nanosystems can be constructed. Utilizing computers and robotic arms, they would be many times smaller than a living cell. For the cryonicist, this means the creation of *nanorobots* capable of being injected into the bloodstream. They would travel through the tissues of the body and eventually enter individual cells, armed with the capability of redirecting that cell's metabolism. Sound way out? Well, viruses do it all the time. A representative repair machine might be one micrometer (micron) wide—the size of a bacterium—and occupy one thousandth the volume of a typical cell.

Once inside the cell, these nanorobots would examine its contents and activity. Sophisticated onboard computers, with more stored data than the cell's entire genome, would assess damage to molecular structures such as nucleic acids and proteins. Then the process of rebuilding would begin. As K. Eric Drexeler, the guru of nanotechnology, states in his book *Engines of Creation*: "By working along molecule by molecule and structure by structure, repair machines will be able to repair whole cells. By working along cell by cell and tissue by tissue they . . . will be able to repair whole organs. . . . Because molecular machines will be able to build molecules and cells from scratch, they will be able to repair even cells damaged to the point of complete inactivity."

Sounds like a cryonicist's dream come true: no matter how sloppy the freeze job, repair systems would make everything right again. Not so fast. First of all, the know-how for such miniaturization is certainly not here yet and may not be for a century or two. Then again, what are a few hundred years of frozen sleep when compared to death and decay—the only alternative? But more important, although most cells and tissues would be able to sustain considerable damage upon cryosuspension, those of the brain could not. The human brain is the most complex structure in the known universe. Intricate patterns of protein molecules stud the surfaces of a hundred billion brain cells and link them into a neutral meshwork of unimaginable complexity. From this meshwork emerges memory, personality, self—the mind. And each mind is unique. No army of repair machines, no matter how sophisticated, would be able to restore the integrity of the human brain once it has been lost. The brain, like Humpty Dumpty, may never be put together again if it falls.

This is one reason why cryosuspension organizations usually offer their clients a choice of whole-body preservation or preservation of the head alone: the body is replaceable; the head is not. Also, freezing only the decapitated head, called *neurosuspension*, is considerably cheaper than whole body suspension. At the Alcor Life Extension Foundation, cryosuspending the entire body costs a minimum of $120,000, as compared to $41,000 for a head job.

Don't have the money? Not even a measly $41,000? Not to worry. You can take out a life insurance policy that names the cryonics institution as beneficiary. When you are declared legally dead, the institution receives the insurance money and you get frozen stiff.

Still not sure you want to go into indefinite cold storage? Perhaps we should take a closer look at exactly how whole-body cryosuspension is performed.

Whole-Body Suspension— A Step-By-Step Account

The following cryopreservation procedures are those of the Alcor Life Extension Foundation. Details might vary elsewhere, but the basic strategies are the same.

1. Legally, a patient cannot be put into cryonic suspension until pronounced clinically dead by an attending physician. If death is anticipated

and not accidental, a team of cryonicists are at
bedside, ready to spring into action. At the moment of death, the body is put on a heart-lung
resuscitator, a mechanical device that adminis-
ters cardiopulmonary resuscitation (CPR) auto-
matically. The purpose of CPR is not to resusci-
tate the patient but rather to keep oxygenated
blood flowing to prevent damage to tissues and
cells. Concurrently, the patient is placed in a tub
of ice and water to begin the cooling-down process. CPR can and should begin as shortly as
two minutes after cessation of breathing and
heartbeat.

2. A combination of about sixteen nutrients and
medications are injected into the patient. They
serve to stabilize the body and prevent any
further *ischemic* damage—injury due to inadequate blood flow. These medications include
antioxidants such as vitamins C and E, anti-
coagulants, pH buffers, and a host of other
drugs. Interestingly, sodium pentobarbitol,
(more popularly known by its trade name, Nem-
butal) or some other sedative is also injected, to
save everyone the embarrassment of having the
patient wake up.

3. Ideally, only ten to fifteen minutes have elapsed
since death occurred. From here on in, things
really begin to heat up—or, more accurately, to
cool down. In a procedure called *femoral cut-
down*, the femoral blood vessels located in the
groin area are surgically accessed and hooked
up to a blood pump/oxygenator. Incorporated
into the blood pump/oxygenator is a high-effi-
ciency blood heat exchanger, which can cool the

patient down to 40° F (15° C) in as little as fifteen minutes. Nauseated yet? Good—let's continue.

4. What happens next depends on where death occurs. If far from Alcor's cryosuspension facility, the patient is first transported to a local mortuary for "blood washout." In this procedure, the patient's circulatory system is flushed free of blood with a special tissue-preserving solution (TPS). The TPS currently used is a commercial preparation called *Viaspan*, routinely employed by hospitals for twenty-four-hour preservation of transplantable organs. The patient is then packed in ice and air-transported to Alcor's perfusion facility.

5. If death occurs close enough to Alcor's home base, procedure 4—washout with TPS—can be eliminated. In this event, the cooled-down body, on circulatory and respiratory support, is whisked immediately to Alcor, where cryoprotective perfusion begins. In an operating room a trained surgeon, using the same techniques employed during bypass surgery, opens the chest cavity and places tubes into the heart and aorta. A heart-lung machine connected to the tubes circulates the cryoprotective agent—a solution of glycerol and sucrose—through the patient's tissues. Over a two-to-four-hour period, approximately 30 percent of the patient's body water is replaced with glycerol. As explained earlier, this is a critical step in avoiding or at least minimizing cell and tissue damage due to ice formation.

6. The patient is disconnected from all machines and tubes, all incisions are closed, and several

days of serious temperature reduction begin. First, the body is placed in a plastic bag and submerged into a silicone oil/dry-ice bath at a temperature of $-108°$ F ($-78°$ C). Cooling the patient to this temperature usually requires thirty-six to forty-eight hours.

7. At this point the patient, who has not eaten for several days, is offered a little snack. After tacitly declining, the patient is wrapped in two pre-cooled sleeping bags, placed in a protective aluminum pod, and lowered into a cooling chamber. Over the ensuing twenty-four hours, liquid nitrogen vapor slowly cools the body to a storage temperature of $-320°$ F ($-196°$ C). Finally, the body is transferred to a dewar, a storage tank filled with liquid nitrogen. Here the body will remain—frozen, lifeless—until revived. There is no sophisticated refrigeration equipment; liquid nitrogen is merely added to the dewar periodically as it evaporates.

In most respects, the procedures involved in neurosuspension are similar to those of whole-body preservation. One major difference is the more demanding surgery of neurosuspension, since blood vessels supplying the head must be accessed from within the chest. After cryoprotective perfusion is completed, the head is surgically removed at the sixth cervical vertebra and then cooled down. Perfusion as well as temperature reduction of the head alone occurs more rapidly than that of the entire body. Bodies of neurosuspension patients are cremated.

Ben Franklin once said he would rather spread his life out so that he lived one year in every hundred years. That way, he could glimpse into the distant future of mankind.

Sounds like Mr. Franklin would have been interested in cryosuspension. A librarian has recently signed up for possible cryoimmortality because she wants to "live long enough to read every book." (If my wife ever decides to cryopreserve herself, I'm sure it will be to live long enough to watch every soap opera.)

Whatever their reasons, people are turning to cryopreservation. Only time will tell whether their dreams of immortality are ever realized and, more important, whether immortality is worth the money and the wait.

Index

Go to Choir
tom morlow